SCADA SECURITY: MACHINE LEARNING CONCEPTS FOR INTRUSION DETECTION AND PREVENTION

Wiley Series On Parallel and Distributed Computing

Series Editor: Albert Y. Zomaya

A complete list of titles in this series appears at the end of this volume.

SCADA SECURITY: MACHINE LEARNING CONCEPTS FOR INTRUSION DETECTION AND PREVENTION

SCADA-BASED IDs SECURITY

Abdulmohsen Almalawi
King Abdulaziz University

Zahir Tari
RMIT University

Adil Fahad
Al Baha University

Xun Yi
RMIT University

This edition first published 2021
© 2021 John Wiley & Sons, Inc.

The right of Abdulmohsen Almalawi, Zahir Tari, Adil Fahad, Xun Yi to be identified as the authors of this work has been asserted in accordance with law.

Registered Office
John Wiley & Sons, Inc., 111 River Street, Hoboken, NJ 07030, USA

Editorial Office
111 River Street, Hoboken, NJ 07030, USA

For details of our global editorial offices, customer services, and more information about Wiley products visit us at www.wiley.com.

Wiley also publishes its books in a variety of electronic formats and by print-on-demand. Some content that appears in standard print versions of this book may not be available in other formats.

Library of Congress Cataloging-in-Publication Data:
Names: Almalawi, Abdulmohsen, author. | Tari, Zahir, author. | Fahad, Adil,
    author. | Yi, Xun, author.
Title: SCADA security : machine learning concepts for intrusion detection
    and prevention / Abdulmohsen Almalawi, King Abdulaziz University, Zahir
    Tari, RMIT University, Adil Fahad, Al Baha University, Xun Yi, Royal
    Melbourne Institute of Technology.
Description: Hoboken, NJ, USA : Wiley, 2021. | Series: Wiley series on
    parallel and distributed computing | Includes bibliographical references
    and index.
Identifiers: LCCN 2020027876 (print) | LCCN 2020027877 (ebook) | ISBN
    9781119606031 (cloth) | ISBN 9781119606079 (adobe pdf) | ISBN
    9781119606352 (epub)
Subjects: LCSH: Supervisory control systems. | Automatic control–Security
    measures. | Intrusion detection systems (Computer security) | Machine
    learning.
Classification: LCC TJ222 .A46 2021 (print) | LCC TJ222 (ebook) | DDC
    629.8/95583–dc23
LC record available at https://lccn.loc.gov/2020027876
LC ebook record available at https://lccn.loc.gov/2020027877

Cover Design: Wiley
Cover Image: © Nostal6ie/Getty Images

Set in 9.5/12.5pt STIXTwoText by SPi Global, Pondicherry, India

Printed in the United States of America

SKY10022785_112320

To our dear parents

CONTENTS

◼◼◼◼◼ FOREWORD

In recent years, SCADA systems have been interfaced with enterprise systems, which therefore exposed them to the vulnerabilities of the Internet and to security threats. Therefore, there has been an increase in cyber intrusions targeting these systems and they are becoming an increasingly global and urgent problem. This is because compromising a SCADA system can lead to large financial losses and serious impact on public safety and the environment. As a countermeasure, Intrusion Detection Systems (IDSs) tailored for SCADA are designed to identify intrusions by comparing observable behavior against suspicious patterns, and to notify administrators by raising intrusion alarms. In the existing literature, there are three types of learning methods that are often adopted by IDS for learning system behavior and building the detection models, namely *supervised, semisupervised*, and *unsupervised*. In supervised learning, anomaly-based IDS requires class labels for both normal and abnormal behavior in order to build normal/abnormal profiles. This type of learning is costly however and time-expensive when identifying the class labels for a large amount of data. Hence, semi-supervised learning is introduced as an alternative solution, where an anomaly-based IDS builds only normal profiles from the normal data that is collected over a period of "normal" operations. However, the main drawback of this learning method is that comprehensive and "purely" normal data are not easy to obtain. This is because the collection of normal data requires that a given system operates under normal conditions for a long time, and intrusive activities may occur during this period of the data collection process. On the another hand, the reliance only on abnormal data for building abnormal profiles is infeasible since the possible abnormal behavior that may occur in the future cannot be known in advance. Alternatively, and for preventing threats that are new or unknown, an anomaly-based IDS uses unsupervised learning methods to build normal/abnormal profiles from unlabeled data, where prior knowledge about normal/abnormal data is not known. Indeed, this is a cost-efficient method since it can learn from unlabeled data. This is because human expertise is not required to identify the behavior (whether normal or abnormal) for each observation in a large amount of training data sets. However, it suffers from low efficiency and poor accuracy.

This book provides the latest research and best practices of unsupervised intrusion detection methods tailored for SCADA systems. In Chapter 3, framework for a SCADA security testbed based on virtualisation technology is described for evaluating and testing the practicality and efficacy of any proposed SCADA security solution. Undoubtedly, the proposed testbed is a salient part for evaluating and testing because the actual SCADA systems cannot be used for such purposes because availability and performance, which are the most important issues, are most likely to be affected when analysing vulnerabilities, threats, and the impact of attacks. In the literature, the k-Nearest Neighbour (k-NN) algorithm was found to be one of top ten most interesting and best algorithms for data mining in general and in particular it has demonstrated promising results in anomaly detection. However, the traditional k-NN algorithm suffers from high and "curse of dimensionality" since it needs a large amount of distance calculations. Chapter 4 describes a novel k-NN algorithm that efficiently works on high-dimensional data of various distributions. In addition, an extensive experimental study and comparison with several algorithms using benchmark data sets were conducted. Chapters 5 and 6 introduce the practicality and possibility of unsupervised intrusion detection methods tailored for SCADA systems, and demonstrate the accuracy of unsupervised anomaly detection methods that build normal/abnormal profiles from unlabeled data. Finally, Chapter 7 describes two authentication protocols to efficiently protect SCADA Systems, and Chapter 8 nicely concludes with the various solutions/methods described in this book with the aim to outline possible future extensions of these described methods.

PREFACE

Supervisory Control and Data Acquisition (SCADA) systems have been integrated to control and monitor industrial processes and our daily critical infrastructures, such as electric power generation, water distribution, and waste water collection systems. This integration adds valuable input to improve the safety of the process and the personnel, as well as to reduce operation costs. However, any disruption to SCADA systems could result in financial disasters or may lead to loss of life in a worst case scenario. Therefore, in the past, such systems were secure by virtue of their isolation and only proprietary hardware and software were used to operate these systems. In other words, these systems were self-contained and totally isolated from the public network (e.g., the Internet). This isolation created the myth that malicious intrusions and attacks from the outside world were not a big concern, and such attacks were expected to come from the inside. Therefore, when developing SCADA protocols, the security of the information system was given no consideration.

In recent years, SCADA systems have begun to shift away from using proprietary and customized hardware and software to using Commercial-Off-The-Shelf (COTS) solutions. This shift has increased their connectivity to the public networks using standard protocols (e.g., TCP/IP). In addition, there is decreased reliance on specific vendors. Undoubtedly, this increases productivity and profitability but will, however, expose these systems to cyber threats. A low percentage of companies carry out security reviews of COTS applications that are being used. While a high percentage of other companies do not perform security assessments, and thus rely only on the vendor reputation or the legal liability agreements, some may have no policies at all regarding the use of COTS solutions.

The adoption of COTS solutions is a time- and cost-efficient means of building SCADA systems. In addition, COST-based devices are intended to operate on traditional Ethernet networks and the TCP/IP stack. This feature allows devices from various vendors to communicate with each other and it also helps to remotely supervise and control critical industrial systems from any place and at any time using the Internet. Moreover, wireless technologies can efficiently be used to provide mobility and local control for multivendor devices at a low cost for installation and maintenance. However, the convergence of state-of-the-art communication technologies exposes SCADA systems to all the inherent vulnerabilities of these technologies.

An awareness of the potential threats to SCADA systems and the need to reduce risk and mitigate vulnerabilities has recently become a hot research topic in the security area. Indeed, the increase of SCADA network traffic makes the manual monitoring and analysis of traffic data by experts time-consuming, infeasible, and very expensive. For this reason, researchers begin to employ Machine Learning (ML)-based methods to develop Intrusion Detection Systems (IDSs) by which normal and abnormal behaviors of network traffic are automatically learned with no or limited domain expert interference. In addition to the acceptance of IDSs as a fundamental piece of security infrastructure in detecting new attacks, they are cost-efficient solutions for minoring network behaviors with high-accuracy performance. Therefore, IDS has been adopted in SCADA systems. The type of information source and detection methods are the salient components that play a major role in developing an IDS. The network traffic and events at system and application levels are examples of information sources. The detection methods are broadly categorized into two types in terms of detection: *signature-based* and *anomaly-based*. The former can detect only an attack whose signature is already known, while the latter can detect unknown attacks by looking for activities that deviate from an expected pattern (or behavior). The differences between the nature and characteristics of traditional IT and SCADA systems have motivated security researchers to develop SCADA-specific IDSs. Recent researches on this topic found that the modelling of measurement and control data, called *SCADA data*, is promising as a means of detecting malicious attacks intended to jeopardize SCADA systems. However, the development of efficient and accurate detection models/methods is still an open research area.

Anomaly-based detection methods can be built by using three modes, namely *supervised*, *semi-supervised*, or *unsupervised*. The class labels must be available for the first mode; however, this type of learning is costly and time-consuming because domain experts are required to label hundreds of thousands of data observations. The second mode is based on the assumption that the training data set represents only one behavior, either normal or abnormal. There are a number of issues pertaining to this mode. The system has to operate for a long time under normal conditions in order to obtain purely normal data that comprehensively represent normal behaviors. However, there is no guarantee that any anomalous activity will occur during the data collection period. On the other hand, it is challenging to obtain a training data set that covers all possible anomalous behaviors that can occur in the future. Alternatively, the unsupervised mode can be the most popular form of anomaly-based detection models that addresses the aforementioned issues, where these models can be built from unlabeled data without prior knowledge about normal/abnormal behaviors. However, the low efficiency and accuracy are challenging issues of this type of learning.

There are books in the market that describe the various SCADA-based unsupervised intrusion detection methods; they are, however, relatively unfocused and lacking much details on the methods for SCADA systems in terms of detection approaches, implementation, data collection, evaluation, and intrusion response. Briefly, this book provides the reader with the tools that are intended to provide practical development and implementation of SCADA security in general. Moreover, this book introduces solutions to practical problems that SCADA intrusion detection systems experience when building unsupervised intrusion detection methods from unlabeled data. The major challenge was to bring various aspects of SCADA intrusion detection systems, such as building unsupervised anomaly detection methods and evaluating their respective performance, under a single umbrella.

The target audience of this book is composed of professionals and researchers working in the field of SCADA security. At the same time, it can be used by researchers who could be interested in SCADA security in general and building SCADA unsupervised intrusion detection systems in particular. Moreover, this book may aid them to gain an overview of a field that is still largely dominated by conference publications and a disparate body of literature.

The book has seven main chapters that are organized as follows. In Chapter 3, the book deals with the establishment of a SCADA security testbed that is a salient part for evaluating and testing the practicality and efficacy of any proposed SCADA security solution. This is because the evaluation and testing using actual SCADA systems are not feasible since their availability and performance are most likely to be affected. Chapter 4 looks in much more detail at the novel efficient k-Nearest Neighbour approach based on Various-Widths Clustering, named kNNVWC, to efficiently address the infeasibility of the use of the k-nearest neighbour approach with large and high-dimensional data. In Chapter 5, a novel SCADA Data-Driven Anomaly Detection (SDAD) approach is described in detail. This chapter demonstrates the practicality of the clustering-based method to extract proximity-based detection rules that comprise a tiny portion compared to the training data, while meanwhile maintain the representative nature of the original data. Chapter 6 looks in detail at a novel promising approach, called GATUD (Global Anomaly Threshold to Unsupervised Detection), that can improve the accuracy of unsupervised anomaly detection approaches that are compliant with the following assumptions: (i) the number of normal observations in the data set vastly outperforms the abnormal observations and (ii) the abnormal observations must be statistically different from normal ones. Finally, Chapter 7 looks at the authentication protocols in SCADA systems, which enable secure communication between all the components of such systems. This chapter describes two efficient TPASS protocols for SCADA systems: one is built on two-phase commitment and has lower computation complexity and the other is based on zero-knowledge proof and has less communication rounds. Both protocols are particularly efficient for the client, who only needs to send a request and receive a response.

ACRONYMS

AGA	American Gas Association
ASCII	American Standard Code for Information Interchange
COTS	Commercial-Off-The-Shelf
CORE	Common Open Research Emulator
CRC	Cyclic Redundancy Check
DDL	Dynamic Link Library
DNP	Distributed Network Protocol
DOS	Denial Of Service
EDMM	Ensemble-based Decision-Making Model
Ek-NN	Exhaustive k-Nearest Neighbor
EMANE	Extendable Mobile Ad-hoc Network Emulator
EPANET	Environmental Protection Agency Network
FEP	Front End Processor
GATUD	Global Anomaly Threshold to Unsupervized Detection
HMI	Human Machine Interface
k-NN	k-Nearest Neighbor
kNNVWC	k-NN based on Various-Widths Clustering
IDS	Intrusion Detection System
IED	Intelligent Electronic Device
IP	Internet Protocol
IT	Information Technology
LAN	Local Area Network
NISCC	National Infrastructure Security Coordination Center
NS2	Network Simulator 2
NS3	Network Simulator 3
OMNET	Objective Modular Network Testbed
OPNET	Optimized Network Engineering Tool
OST	Orthogonal Structure Tree
OSVDB	Open Source Vulnerability DataBase
PCA	Principal Component Analysis
PLC	Programmable Logic Controller
PLS	Partial Least Squares
RTU	Remote Terminal Unit
SCADA	Supervisory Control And Data Acquisition
SCADAVT	SCADA security testbed based on Virtualization Technology

SDAD	SCADA Data-driven Anomaly Detection
TCP	Transmission Control Protocol
TPASS	Threshold Password-Authenticated Secret S in the boo.. It is haring
UDP	User Datagram Protocol
USB	Universal Serial Bus

CHAPTER 1

Introduction

This aim of this introductory chapter is to motivate the extensive research work carried in this book, highlighting the existing solutions and their limitations, and putting
in context the innovative work and ideas described in this book.

1.1 OVERVIEW

Supervisory Control and Data Acquisition (SCADA) systems have been integrated
to control and monitor industrial processes and our daily critical infrastructures such
as electric power generation, water distribution and waste water collection systems.
This integration adds valuable input to improve the safety of the process and the
personnel and to reduce operation costs (Boyer, 2009). However, any disruption to
SCADA systems can result in financial disasters or may lead to loss of life in a worst
case scenario. Therefore, in the past, such systems were secure by virtue of their isolation and only proprietary hardware and software were used to operate these systems.
In other words, these systems were self-contained and totally isolated from the public
network (e.g., the Internet). This isolation created the myth that malicious intrusions
and attacks from the outside world were not a big concern and that such attacks were
expected to come from the inside. Therefore, when developing SCADA protocols,
the security of the information system was given no consideration.

In recent years, SCADA systems have begun to shift away from using proprietary
and customized hardware and software to using Commercial-Off-The-Shelf (COTS)
solutions. This shift has increased their connectivity to the public networks using
standard protocols (e.g., TCP/IP). In addition, there is decreased reliance on a single
vendor. Undoubtedly, this increases productivity and profitability but will, however,
expose these systems to cyber threats (Oman et al., 2000). According to a survey
published by the SANS Institute (Bird and Kim, 2012), only 14% of organizations
carry out security reviews of COTS applications that are being used, while over 50%
of other organizations do not perform security assessments and rely only on vendor
reputation or the legal liability agreements, or they have no policies at all regarding
the use of COTS solutions.

The adoption of COTS solutions is a time- and cost-efficient means of building SCADA systems. In addition, COST-based devices are intended to operate on

SCADA Security: Machine Learning Concepts for Intrusion Detection and Prevention,
First Edition. Abdulmohsen Almalawi, Zahir Tari, Adil Fahad and Xun Yi.
© 2021 John Wiley & Sons, Inc. Published 2021 by John Wiley & Sons, Inc.

traditional Ethernet networks and the TCP/IP stack. This feature allows devices from various vendors to communicate with each other, and also helps to remotely supervise and control critical industrial systems from any place and at any time using the Internet. Moreover, wireless technologies can efficiently be used to provide mobility and local control for multivendor devices at a low cost for installation and maintenance. However, the convergence of state-of-the-art communication technologies exposes SCADA systems to all the inherent vulnerabilities of these technologies. In what follows, we discuss how the potential cyber-attacks against traditional IT can also be possible against SCADA systems.

- **Denial of Services (DoS) attacks.** This is a potential attack on any Internet-connected device where a large number of spurious packets are sent to a victim in order to consume excessive amounts of endpoint network bandwidth. A packet flooding attack (Houle et al., 2001) is often used as another term for a DoS attack. This type of attack delays or totally prevents the victim from receiving the legitimate packets (Householder et al., 2001). SCADA networking devices that are exposed to the Internet such as routers, gateways and firewalls are susceptible to this type of attack. Long et al. (2005) proposed two models of DoS attacks on a SCADA network using reliable simulation. The first model was directly launched to an endpoint (e.g., controller or a customer-edge router connecting to the Internet), while the second model is an indirect attack, where the DoS attack is launched on a router (on the Internet) that is located in the path between the plant and endpoint. In this study, it was found that DoS attacks that were launched directly (or indirectly) cause excessive packet losses. Consequently, a controller that receives the measurement and control data late or not at all from the devices deployed in the field will make a decision based on old data.

- **Propagation of malicious codes.** Such types of attack can occur in various forms such as viruses, Trojan horses, and worms. They are potential threats to SCADA systems that are directly (or indirectly) connected to the Internet. Unlike worms, viruses and Trojans require a human action to be initiated. However, all these threats are highly likely as long as the personnel are connected to the Internet through the corporate network, which is directly connected to the SCADA system, or if they are allowed to plug their personal USBs into the corporate workstations. Therefore, a user can be deceived into downloading a contaminated file containing a virus or installing software that appears to be useful. Shamoon (Bronk and Tikk-Ringas, 2013), Stuxnet (Falliere et al., 2011), Duqu (Bencsáth et al., 2012), and Flame (Munro, 2012) are examples of such threats targeting SCADA systems and oil and energy sectors.

- **Inside threats.** The employees who are disgruntled or intend to divulge valuable information for malicious reasons can pose real threats and risks that should be taken seriously. This is because employees usually have unrestricted access to the SCADA systems and also know the configuration settings of these systems. For instance, the attack on the sewage treatment system in Maroochy Shire, South-East Queensland (Australia) in 2001 (Slay and Miller, 2007) is an

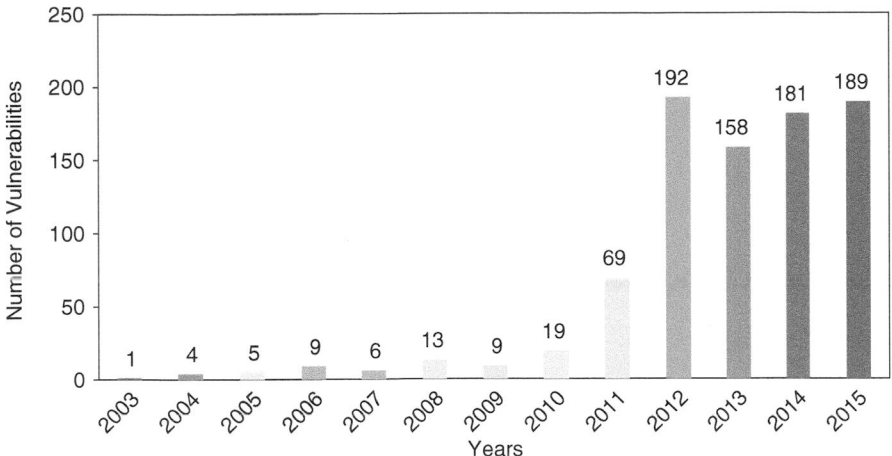

Figure 1.1 SCADA vulnerabilities revealed since 2001 in OSVDB.

example of an attack that was launched by a disgruntled employee, where the attacker took over the control devices of a SCADA system and caused 800,000 litres of raw sewage to spill out into local parks and rivers.

- **Unpatched vulnerabilities.** The existence of vulnerabilities is highly expected in any system and it is known that hackers always exploit unpatched vulnerabilities to obtain access and to control the targeted system. Even though the vendors immediately release the patches for the identified vulnerabilities, it is challenging to install these patches on SCADA systems that run twenty-four-by-seven. Therefore, such systems will remain vulnerable for weeks or months. As depicted in Figure 1.1, and according to the independent and Open Source Vulnerability DataBase (OSVDB)[1] for the security community, vulnerabilities targeting SCADA systems have substantially increased over the past three years since 2011.

- **Nontechnical (social engineering) attacks.** This type of attack can bypass state-of-the-art security technologies that cost millions of dollars. In general, the attackers initially try to obtain sensitive information such as the design, operations, or security controls of the targeted SCADA system. There are a number of ways to gather such information. If the network access credentials of ex-employees are not immediately disabled, they can be revealed to another party in order to profit from the information, or as a desire for revenge. In another way, such critical information can be easily obtained from current employees as long as they are known by building a trust relationship or by knowing some information about a naive employee who is allowed to remotely control and monitor the systems via the Internet, all of which can help the attacker to answer the expected questions when calling up the central office

[1]http://osvdb.org/

to tell them that s/he forgot the network access credentials and assistance is needed to connect to the field network.

The security concepts that have been extensively used in traditional IT systems (e.g., *management, filtering, encryption,* and *intrusion detection*) can be adapted to mitigate the risk of the aforementioned potential threats against SCADA systems. However, these concepts cannot be directly applied without considering the *nature* of SCADA systems. For instance, the resource constraints of SCADA devices, such as low bandwidth, processing power, and memory, complicate the integration of complex cryptography, especially with legacy devices. All the SCADA protocols were developed without any consideration given to information security and, therefore, they lack authentication and integrity. Two solutions to secure the SCADA communications are: placing the cryptographic technologies at each end of the communication medium (American Gas Association (AGA), 2006; Tsang and Smith, 2008), or directly integrating them into the protocol, such as a secure DNP3 that protects the communication between master stations and outstations such as PLCs, RTUs, and IEDs (Majdalawieh et al., 2006).

Apart from the efforts to authenticate and encrypt SCADA communication links, it is still an ***open research challenge*** to secure the tens of SCADA protocols that are being used or to develop security modules to protect the communication link between two parties. AGA (American Gas Association (AGA), 2006) highlighted the challenges in building security modules that can be broadly summarized into two points: (i) the additional latency can be introduced by a secure protocol and (ii) the sophisticated key management system requires high bandwidth and additional communication channels that SCADA communication links are lacking.

Similarly, the traffic filtering process between a SCADA network and a corporate network using firewalls is a considerable countermeasure to mitigate the potential threats. However, although modern firewalls are efficient for analysing traditional IT traffic, they are incapable of in-depth analysis of the SCADA protocols. To design firewalls tailored to SCADA systems, the UK governments National Infrastructure Security Co-ordination Center (NISCC) published its guidelines for the appropriate use of firewalls in SCADA networks (Byres et al., 2005). It was proposed that a microfirewall should be embedded within each SCADA device to allow only the traffic relevant to the host devices. However, the computational power of SCADA devices can be a challenging issue to support this type of firewall.

Firewalls can be configured using restrict-constrained rules to control traffic in and out of the SCADA network; however, this will conflict with the feature allowing remote maintenance and operation by vendors and operators. Additionally, firewalls are assumed to be physically placed between the communication endpoints to examine each packet prior to passing it to the receiver. This may cause a latency that is not acceptable in real-time networks. Since firewalls do not know the "normal" operational behavior of the targeted system, they cannot stop malicious control messages, which may drive the targeted system from its expected and normal behavior, when they are sent from a compromised unit that is often used to remotely control and monitor SCADA networks. Moreover, it is beyond the ability of firewalls when the attacks are initiated internally using an already-implanted malicious code or directly by an employee. Stuxnet (Falliere et al., 2011), Duqu (Bencsáth et al., 2012), and Flame (Munro, 2012) are the recent cyber-attacks that were initiated from inside automation systems. Therefore, the reliance only on firewalls is not sufficient to mitigate the potential threats to SCADA systems. Hence, an additional defense needs to be installed to monitor already predefined (or unexpected) patterns for either network traffic or system behavior in order to detect any intrusion attempt. The system using such a method is known in the information security area as an *Intrusion Detection System* (IDS).

There is no security countermeasures that can completely protect the target systems from potential threats, although a number of countermeasures can be used in conjunction with each other in order to build a robust security system. An IDS (Intrusion Detection System) is one of the security methods that has demonstrated promising results in detecting malicious activities in traditional IT systems. The source of audit data and the detection methods are the main, salient parts in the development of an IDS. The network traffic, system-level events and application-level activities are the most usual sources of audit data. The detection methods are categorized into two strategies: *signature-based* and *anomaly-based*. The former searches for an attack whose signature is already known, while the latter searches for activities that deviate from an expected pattern or from the predefined normal behavior.

Due to the differences between the nature and characteristics of traditional IT and SCADA systems, there has been a need for the development of SCADA-specific IDSs, and in recent years this has become an interesting research area. In the literature, they vary in terms of the information source being used and in the analysis strategy. Some of them use SCADA network traffic (Linda et al., 2009; Cheung et al., 2007; Valdes and Cheung, 2009), system-level events (Yang et al., 2006), or measurement and control data (values of sensors and actuators) (Rrushi et al., 2009b; Fovino et al., 2010a, 2012; Carcano et al., 2011) as the information source to detect malicious, uncommon or inappropriate actions of the monitored system using various analysis strategies which can be signature-based, anomaly-based or a combination of both.

It is believed that modeling of measurement and control data is a promising means of detecting malicious attacks intended to jeopardize a targeted SCADA system. For instance, the Stuxnet worm is a sophisticated attack that targets a control system and initially cannot be detected by the antivirus software that was installed in the victim (Falliere et al., 2011). This is because it used zero-day vulnerabilities and validated its drivers with trusted stolen certificates. Moreover, it could hide its modifications using sophisticated PLC rootkits. However, the final goal of this attack cannot be hidden since the manipulation of measurement and control data will make the behavior of the targeted system deviate from previously seen ones. This is the **main motivation of this book**, namely to explain in detail *how to design SCADA-specific IDSs using SCADA data (measurement and control data)*, thus enabling the reader to build/implement an information source that monitors the internal behavior of a given system and protects it from malicious actions that are intended to sabotage or disturb the proper functionality of the targeted system.

As previously indicated, the analysis/modeling method, which will be used to build the detection model using SCADA data, is the second most important part after the selection of the information source when designing an Intrusion Detection System (IDS). It is difficult to build the "normal" behavior of a given system using observations of the raw SCADA data because, firstly, it cannot be guaranteed that all observations represent one behavior as either "*normal*" or "*abnormal*", and therefore domain experts are required for the labeling of each observation, and this process is prohibitively expensive; secondly, in order to obtain purely "normal" observations that comprehensively represent "normal" behavior, this requires a given system to be run for a long period under normal conditions, and this not practical; and, finally, it is challenging to obtain observations that will cover all possible abnormal behavior that can occur in the future. Therefore, we strongly argue that the design of a SCADA-specific IDS that uses **SCADA data** as well as **operating in unsupervised mode**, where the labeled data is not available, has great potential as a means of addressing the aforementioned issues. The unsupervised IDS can be a time- and cost-efficient means of building detection models from unlabeled data; however, this requires an efficient and accurate method to differentiate between the normal and abnormal observations without the involvement of experts, which is costly and prone to human error. Then, from observations of each behavior, either normal or abnormal, the detection models can be built.

1.2 EXISTING SOLUTIONS

A layered defense could be the best security mechanism, where each layer in the computer and network system is provided with a particular security countermeasure. For instance, organizations deploy firewalls between their private networks and others to prevent unauthorized users from entering. However, firewalls cannot address all risks and vulnerabilities. Therefore, an additional security layer is required. The last component at the security level is the IDS, which is used to monitor intrusive activities (Pathan, 2014). The concept of an IDS is based on the assumption that the behavior of intrusive activities are noticeably distinguishable from the normal ones (Denning, 1987). Since the last decade, compared to other security countermeasures, the deployment of IDS technology has attracted great interest from the traditional IT systems domain (Pathan, 2014). The promising functionalities of this technology have encouraged researchers and practitioners concerned with the security of SCADA systems to adopt this technology while taking into account the nature and characteristics of SCADA systems.

To design an IDS, two main processes are often considered: first, the selection of the information source (e.g., network-based, application-based) to be used, through which anomalies can be detected; second, the building of the detection models using the specified information source. SCADA-specific IDSs can be broadly grouped into three categories in terms of the latter process: *signature-based detection* (Digitalbond, 2013), *anomaly detection* (Linda et al., 2009; Kumar et al., 2007; Valdes and Cheung, 2009; Yang et al., 2006; Ning et al., 2002; Gross et al., 2004), and *specification-based detection* (Cheung et al., 2007; Carcano et al., 2011; Fovino et al., 2010a; Fernandez et al., 2009). Recently, several signature-based rules (Digitalbond, 2013) have been designed to specifically detect particular attacks on SCADA protocols. The rules can perfectly detect *known attacks* at the SCADA network level. To detect *unknown attacks* at the SCADA network level, a number of methods have been proposed. Linda et al. (2009) suggested a window-based feature extraction method to extract important features of SCADA network traffic and then used a feed-forward neural network with the back propagation training algorithm for modeling the boundaries of normal behavior. However, this method suffers from the great amount of execution time required in the training phase, in addition to the need for relearning the boundaries of normal behavior upon receiving new behavior.

The model-based detection method proposed in Valdes and Cheung (2009) illustrates communication patterns. This is based on the assumption that the communication patterns of control systems are regular and predictable because SCADA has specific services as well as interconnected and communicated devices that are already predefined. This method is useful in providing a border monitoring of the requested services sand devices. Similarly, Gross et al. (2004) proposed a collaborative method, named "selecticast", which uses a centralized server to disperse among ID sensors any information about activities coming from suspicious IPs. Ning et al. (2002) identify causal relationships between alerts using prerequisites and consequences. In essence, these methods fail to detect *high-level control attacks*, which are the most difficult threats to combat successfully (Wei et al., 2011). Furthermore, SCADA network level methods are not concerned with the operational meaning of the process parameter values, which are carried by SCADA protocols, as long as they are not violating the specifications of the protocol being used or a broader picture of the monitored system.

Thus, analytical models based on the full system's specifications have been suggested in the literature. Fovino et al. (2010a) proposed an analytical method to identify critical states for specific-correlated process parameters. Therefore, the developed detection models are used to detect malicious actions (such as high-level control attacks) that try to drive the targeted system into a critical state. In the same direction, Carcano et al. (2011) and Fovino et al. (2012) extended this idea by identifying critical states for specific-correlated process parameters. Then, each critical state is represented by a multivariate vector, each vector being a reference point to measure the degree of criticality of the current system. For example, when the distance of the current system state is close to any critical state, it shows that the system is approaching a critical state. However, the critical state-based methods require full specifications of all correlated process parameters in addition to their respective acceptable values. Moreover, the analytical identification of critical states for a relatively large number of correlated process parameters is time-expensive and difficult. This is because the complexity of the interrelationship among these parameters is proportional to their numbers. Furthermore, any change in the system brought about by adding or removing process parameters will require the same effort again. Obviously, human errors are highly expected in the identification process of critical system states.

Due to the aforementioned issues relating to analytical methods, SCADA data-driven methods have been proposed to capture the mechanistic behavior of SCADA systems without a knowledge of the physical behavior of the systems. It was experimentally found by Wenxian and Jiesheng (2011) that operational SCADA data for wind turbine systems are useful if they are properly analyzed to indicate the condition of the system that is being supervised. A number of SCADA data-driven methods for anomaly detection have appeared in the literature. Jin et al. (2006) extended the set of invariant models by *a value range model* to detect anomalous values in the values for a particular process parameter. A predetermined threshold is proposed for each parameter and any value exceeding this threshold is considered as anomalous. This method can detect the anomalous values of an individual process parameter. However, the value of an individual process parameter may not be abnormal, but, in combination with other process parameters, may produce abnormal observation, which very rarely occurs. These types of parameter are called *multivariate parameters* and are assumed to be directly (or indirectly) correlated. Rrushi et al. (2009b) applied probabilistic models to estimate the normalcy of the evolution of values of multivariate process parameters. Similarly, Marton et al. (2013) proposed a data-driven method to detect abnormal behaviour in industrial equipment, where two multivariate analysis methods, namely principal component analysis (PCA) and partial least squares (PLS), are combined to build the detection models. Neural network-based methods have been proposed to model the normal behavior for various SCADA applications. For instance, Gao et al. (2010) proposed a neural-network-based intrusion detection system for water tank control systems. In a different application, this method has been adapted by Zaher et al. (2009) to build the normal behaviour for a wind turbine to identify faults or unexpected behavior (anomalies).

Although the results for the aforementioned SCADA data-driven methods are promising, they work only in supervised or semisupervised modes. The former method is applicable when the labels for both normal/abnormal behavior are available. Domain experts need to be involved in the labeling process but it is costly and time-consuming to label hundreds of thousands of data observations (instances). In addition, it is difficult to obtain abnormal observations that comprehensively represent anomalous behavior, while in the latter mode a one-class problem (either normal or abnormal data) is required to train the model. Obtaining a normal training data set can be done by running a target system under normal conditions and the collected data is assumed to be normal. To obtain purely normal data that comprehensively represent normal behavior, the system has to operate for a long time under normal conditions. However, this cannot be guaranteed and therefore any anomalous activity occurring during this period will be learned as normal. On the other hand, it is challenging to obtain a training data set that covers all possible anomalous behavior that can occur in the future.

Unlike supervised, semisupervised, and analytical solutions, this book is about **designing unsupervised anomaly detection methods**, where experts are not required to prepare a labeled training data set or analytically define the boundaries of normal/abnormal behavior of a given system. In other words, this book is interested in developing a **robust unsupervised intrusion detection system** that automatically identifies, from unlabeled SCADA data, both normal and abnormal behavior, and then extracts the proximity-detection rules for each behavior.

1.3 SIGNIFICANT RESEARCH PROBLEMS

In recent years, many researchers and practitioners have turned their attention to SCADA data to build data-driven methods that are able to learn the *mechanistic* behavior of SCADA systems without a knowledge of the physical behavior of these systems. Such methods have shown a promising ability to detect anomalies, malfunctions, or faults in SCADA components. Nonetheless, it remains a relatively **open research area** to develop unsupervised SCADA data-driven detection methods that can be time- and cost-efficient for learning detection methods from unlabeled data. However, such methods often have a low detection accuracy. The focus of this book is about the **design of an efficient and accurate unsupervised SCADA data-driven IDS**, and four main research problems are formulated here for this purpose. Three of these pertain to the development of methods that are used to build a robust unsupervised SCADA data-driven IDS. The fourth research problem relates to the design of a framework for a SCADA security testbed that is intended to be an evaluation and testing environment for SCADA security in general and for the proposed unsupervised IDS in particular.

1. **How to design a SCADA-based testbed that is a realistic alternative for real SCADA systems so that it can be used for proper SCADA security evaluation and testing purposes.** An evaluation of the security solutions of SCADA systems is important. However, actual SCADA systems cannot be used for such a purpose because availability and performance, which are the most important issues, are most likely to be affected when analysing vulnerabilities, threats, and the impact of attacks. To address this problem, "real SCADA testbeds" have been set up for evaluation purposes, but they are costly and beyond the reach of most researchers. Similarly, small real SCADA testbeds have also been set up; however, they are still proprietary and location-constrained. Unfortunately, such labs are not available to researchers and practionners interested in working on SCADA security. Hence, the design of a SCADA-based testbed for that purpose will be very useful for evaluation and testing purposes. Two essential parts could be considered here: *SCADA system components* and a *controlled environment*. In the former, both high-level and field-level components will be considered and the integration of a real SCADA protocol will be devised to realistically produce SCADA network traffic. In the latter, it is important to model a controlled environment such as smart grid power or water distribution systems so that we can produce realistic SCADA data.

2. **How to make an existing suitable data mining method deal with large high-dimensional data.** Due to the specific nature of the unsupervised SCADA

systems, an IDS will be designed here based on SCADA data-driven methods from the unlabeled SCADA data which, it is highly expected, will contain anomalous data; the task is intended to give an anomaly score for each observation. The k-Nearest Neighbour (k-NN) algorithm was found, from an extensive literature review, to be one of the top ten most interesting and best algorithms for data mining in general (Wu et al., 2008), and, in particular, it has demonstrated promising results in anomaly detection (Chandola et al., 2009). This is because the anomalous observation is assumed to have a neighborhood in which it will stand out, while a normal observation will have a neighborhood where all its neighbors will be exactly like it. However, having to examine all observations in a data set in order to find k-NN for an observation x is the main drawback of this method, especially with a vast amount of high dimensional data. To efficiently utilize this method, the reduction of computation time in finding k-NN is the aim of this research problem that this book endeavors to address.

3. **How to learn clustering-based proximity rules from unlabeled SCADA data for SCADA anomaly detection methods.** To build efficient SCADA data-driven detection methods, the efficient proposed k-NN algorithm in problem 2 is used to assign an anomaly score to each observation in the training data set. However, it is impractical to use all the training data in the anomaly detection phase. This is because a large memory capacity is needed to store all scored observations and it is computationally infeasible to compute the similarity between these observations and each current new observation. Therefore, it would be ideal to efficiently separate the observations, which are highly expected to be consistent (normal) or inconsistent (abnormal). Then, a few proximity detection rules for each behavior, whether consistent or inconsistent, are automatically extracted from the observations that belong to that behavior.

4. **How to compute a global and efficient anomaly threshold for unsupervised detection methods.** Anomaly-scoring-based and clustering-based methods are among the best-known ones that are often used to identify the anomalies in unlabeled data. With anomaly-scoring-based methods (Eskin et al., 2002; Angiulli and Pizzuti, 2002; Zhang and Wang, 2006), all observations in a data set are given an anomaly score and therefore actual anomalies are assumed to have the highest scores. The key problem is how to find the near-optimal cut-off threshold that minimizes the false positive rate while maximizing the detection rate. On the one hand, clustering-based methods (Portnoy et al., 2001; Mahoney and Chan, 2003a; Portnoy et al., 2001; Jianliang et al., 2009; Münz et al., 2007) group similar observations together into a number of clusters, and anomalies are identified by making use of the fact that those anomalous observations will be considered as outliers, and therefore will not be assigned to any cluster, or they will be grouped in small clusters that have some characteristics that are different from those of normal clusters. However, the detection of anomalies is controlled through several parameter choices within each used detection method. For instance, given the top 50% of the observations that have the highest anomaly scores, these are assumed as anomalies. In this case, both

detection and false positive rates will be much higher. Similarly, labeling a low percentage of largest clusters as normal in clustering-based intrusion detection methods will result in higher detection and false positive rates. Therefore, the effectiveness of unsupervised intrusion methods is sensitive to parameter choices, especially when the boundaries between normal and abnormal behavior are not clearly distinguishable. Thus, it would be interesting to identify the observations whose anomaly scores are extreme and significantly deviate from others, and then such observations are assumed to be "abnormal". On another hand, the observations whose anomaly scores are significantly distant from "abnormal" ones will be assumed to be "normal". Then, the ensemble-based supervised learning is proposed to find a global and efficient anomaly threshold using the information of both "normal"/"abnormal" behavior.

1.4 BOOK FOCUS

This section summarizes the important lessons learned from the development of robust unsupervised SCADA data-driven Intrusion Detection Systems (IDSs), which are detailed in the various chapters of this book. The first lesson relates to the design of a SCADA security testbed through which the practicality and efficiency of SCADA security solutions are evaluated and tested, while, the remaining three aspects focus on the details of the various elements of a robust unsupervised SCADA data-driven IDS.

- The evaluation and testing of security solutions tailored to SCADA systems is a challenging issue facing researchers and practitioners working on such systems. Several reasons for this include: privacy, security, and legal constraints that prevent organizations from publishing their respective SCADA data. In addition, it is not feasible to conduct experiments on actual live systems, as this is highly likely to affect their availability and performance. Moreover, the establishment of a real SCADA Lab can be costly and place-constrained, and therefore unavailable to all researchers and practitioners. In this book, a framework for a SCADA security testbed is described to build a full SCADA system based on a hybrid of emulation and simulation methods. A real SCADA protocol is implemented and therefore realistic SCADA network traffic is generated. Moreover, a key benefit of this framework is that it is a realistic alternative to real-world SCADA systems and, in particular, it can be used to evaluate the accuracy and efficiency of unsupervised SCADA data-driven Intrusion Detection Systems (IDSs).
- Unsupervised learning for anomaly-detection methods is time- and cost-efficient since they can learn from unlabeled data. This is because human expertise is not required to identify the behavior (whether normal or abnormal) for each observation in a large amount of training data sets. Anomaly scoring methods are believed to be promising automatic methods for assigning an anomaly degree to each observation (Chandola et al., 2009). The k-NN method

is one of the most interesting and best methods for computing the degree of anomaly based on neighborhood density of a particular observation (Wu et al., 2008). However, this method requires high computational cost, especially with large and high-dimensional data that we expect to have in the development of an unsupervised SCADA data-driven IDS. Therefore, this book describes an efficient *k*-nearest neighbor-based method, called *k*NNVWC (*k*-Nearest Neighbor approach based on Various-Widths Clustering), which utilizes a novel various-width clustering algorithm and triangle inequality.

- It is not feasible to retain all the training data in SCADA data-driven anomaly detection methods, especially when these are built from a large training data set. This is because such detection methods will be used for on-line monitoring, and therefore the more information retained in the detection methods, the larger the memory capacity required and the higher the computation cost required. To address this issue, this book describes a clustering-based method to extract proximity-based detection rules, called SDAD (SCADA Data-Driven Anomaly Detection), which are assumed to be a tiny portion compared to the training data, for each behavior (normal and abnormal). Each rule comprehensively represents a subset of observations that represent only one behavior.

- Unsupervised learning for anomaly-detection methods are based mainly on assumptions to find the near-optimal anomaly detection threshold. Therefore, the accuracy of the detection methods is based on the validity of the assumptions. This book, however, describes an efficient method, called GATUD (Global Anomaly Threshold to Unsupervised Detection), which firstly identifies observations whose anomaly scores significantly deviate from others to represent "abnormal" behavior. On the other hand, a tiny portion of observations whose anomaly scores are the smallest are considered to represent "normal" behavior. Then an ensemble-based decision-making method is described, which aims to find a global and efficient anomaly threshold using the information of both "normal"/"abnormal" behavior.

1.5 BOOK ORGANIZATION

The remainder of the book is structured as follows. Chapter 2 gives an introduction to readers who do not have an understanding of SCADA systems and their architectures, and the main components. This includes a description of the relationship between the main components and three generations of SCADA systems. The classification of a SCADA IDS based on its architecture and implementation is described.

Chapter 3 describes in detail SCADAVT, a framework for a SCADA security testbed based on virtualization technology. This framework is used to create a simulation of the main SCADA system components and a controlled environment. The main SCADA components and real SCADA protocol (e.g., Modbus/TCP) are integrated. In addition, a server, which acts as a surrogate for water distribution systems, is introduced. This framework is used throughout the book to simulate a realistic SCADA system for supervising and controlling a water distribution system. This simulation is mentioned in the other chapters to evaluate and test anomaly detection models for SCADA systems.

Chapter 4 describes in detail kNNVWC, an efficient method that finds the k-nearest neighbors in large and high-dimensional data. In kNNVWC, a new various-widths clustering algorithm is introduced, where the data is partitioned into a number of clusters using various widths. Triangle inequality is adapted to prune unlikely clusters in the search process of k-nearest neighbors for an observation. Experimental results show that kNNVWC performs well in finding k-nearest neighbors compared to a number of k-nearest neighbor-based algorithms, especially for a data set with high dimensions, various distributions, and large size.

Chapter 5 describes SDAD, a method that extracts proximity-based detection rules from unlabeled SCADA data, based on a clustering-based method. The evaluation of SDAD is carried out using real and simulated data sets. The extracted proximity-based detection rules show a significant detection accuracy rate compared with an existing clustering-based intrusion detection algorithm.

Chapter 6 describes GATUD, a method that finds a global and efficient anomaly threshold. GATUD is proposed as an add-on component that can be attached to any unsupervised anomaly detection method in order to define the near-optimal anomaly threshold. GATUD shows significant and promising results with two unsupervised anomaly detection methods.

Chapter 7 looks at the authentication aspects related to SCADA environments. It describes two innovative protocols which are based on TPASS (Threshold Password-Authenticated Secret Sharing) protocols; one is built on two-phase commitment and has lower computation complexity and the other is based on zero-knowledge proof and has less communication rounds. Both protocols are particularly efficient for the client, who only needs to send a request and receive a response. Additionally, this chapter provides rigorous proofs of security for the protocols in the standard model.

Finally, Chapter 8 concludes with a summary of the various tools and methods described in this book to the extant body of research and suggests possible directions for future research.

CHAPTER 2

Background

This chapter provides the readers with the necessary background to understand the various elements of this book. This includes an introduction to SCADA systems and their architectures and main components. In addition, the description of the relationship between the main components and three generations of SCADA systems are introduced. The classification of a SCADA-based Intrusion Detection System (IDS) based on its architecture and implementation are also described.

2.1 SCADA SYSTEMS

SCADA (Supervisory Control And Data Acquisition) is an important computer-controlled industrial system that continuously monitors and controls many different sections of industrial infrastructures such as oil refineries, water treatment and distribution systems, and electric power generation plants, to name a few. A SCADA system is responsible for supervising and monitoring industrial and infrastructure processes by gathering measurements and control data from the deployed field devices at the field level. The collected data are then sent to a central site for further processing and analysis. The information and status of the supervised and monitored processes can be displayed on a humanmachine interface (HMI) at the home station in a logical and organized fashion. If an abnormal event occurs, the operators can analyse the gathered data and put in place the necessary controls. Because these industrial systems are large and distributed complexes, it is necessary to continuously and remotely monitor and control different sections of the plant in order to ensure its proper operation by a central master unit.

2.1.1 Main Components

SCADA provides the facility of continuously supervising and controlling the industrial plant or process equipment. The main components of a typical SCADA system include the Master Terminal Unit (MTU), Programmable Logic Controller (PLC), Remote Terminal Unit (RTU), Communication Media, and Human–Machine Interface (HMI).

SCADA Security: Machine Learning Concepts for Intrusion Detection and Prevention,
First Edition. Abdulmohsen Almalawi, Zahir Tari, Adil Fahad and Xun Yi.
© 2021 John Wiley & Sons, Inc. Published 2021 by John Wiley & Sons, Inc.

- **MTU** is the core of a SCADA system that gathers the information from the distributed RTUs and analyses this information for the control process. The plant performance is evaluated through histogram generation, standard deviation calculation, plotting one parameter with respect to another, and so on. Based on the performance analysis, an operator may decide to monitor any channel more frequently, change the limits, shut down the terminal units, and so on. The software can be designed according to the applications and the type of analysis required. The human operator sometimes cannot find the best operating policy for a plant that will minimise the operating costs. Because of this deficiency caused by the enormous complexity of a typical process plant, the master computer station with a high speed and the programmed intelligence of the digital computer are used to analyse the situation and find out the best policy. The MTU monitors, controls, and coordinates the activities of various RTUs and sends supervisory control commands to the process plant.

- **Field devices (RTUs, PLCs, and IEDs)** are computer-based components, that are deployed at a remote site to gather data from sensors and actuators. Each field device may be connected to one (or more) sensors and actuators that are directly connected to physical equipment such as pumps, valves, motors, etc. The main function of such devices is to convert the electrical signals coming from sensors and actuators into digital values in order to be sent to the MTU for further processing and analysis using a communication protocol (e.g. Modbus). On the another hand, they can convert a digital command message, which is received from the MTU, into an electrical signal in order to control actuators that are being supervised and controlled. Even though these field-level devices, RTUs, PLCs, and IEDs, are intended to be deployed at a remote site, they have different functionalities. RTUs collect data from sensors and send it back to the MTU and then the MTU takes a decision based on the this data and sends a command to the actuators. In addition to the same function of RTUs, PLCS can collect data from sensors and, based on the collected data, can send commands to actuators. That is, PLCs can process the data locally and take the decision without contacting the MTU. IEDs are part of control systems such as transformers, circuit breakers, sensors, etc., and can be controlled via PLCs or RTUs.

- **HMI** provides an efficient human–machine interface through which the operator can monitor and control the end devices such as sensors and actuators. That is, the information of the current state of the supervised and controlled process can be graphically displayed to the user, and therefore s/he can be updated with alerts, warnings, and urgent messages. In addition, HMI allows the user to entirely interact with the system.

- **Historian** is a database that is used to store all data gathered from the system, such as measurement and control data, events, alarms, operator's activities, etc. These data are used for historical, auditing, and analysis purposes.

2.1.2 Architecture

A SCADA network provides the communication infrastructure for different field devices, such as PLCs and RTUs on a plant. These field devices are remotely monitored and controlled throughout the SCADA network. To make the network communication more efficient and secure, many modern computing technologies have evolved from a monolithic system to a distributed system and to a current networked system.

Monolithic systems (*First Generation*). Such systems are considered to be the first-generation SCADA systems. At that time, the concept of networks were nonexistent in general, and therefore SCADA systems were deployed as stand-alone systems and there was no connectivity to other systems. Figure 2.1 illustrates the typical architecture of this generation. Typically, a SCADA master uses Wide Area Networks (WANs) to communicate with field devices using communication protocols that were developed by vendors of field devices. In addition, these protocols had limited functionality and they could only do scanning and controlling points within RTUs. The communication between the master and field devices (e.g. RTUs) were performed at the bus level using a proprietary adapter. To avoid a system's failure, two identically equipped mainframe systems are used, one to be a primary with another as backup. The latter will take over when failure of the primary is detected.

Distributed systems (*Second Generation*). Figure 2.2 depicts a typical second-generation SCADA architecture. With the development of Local Area

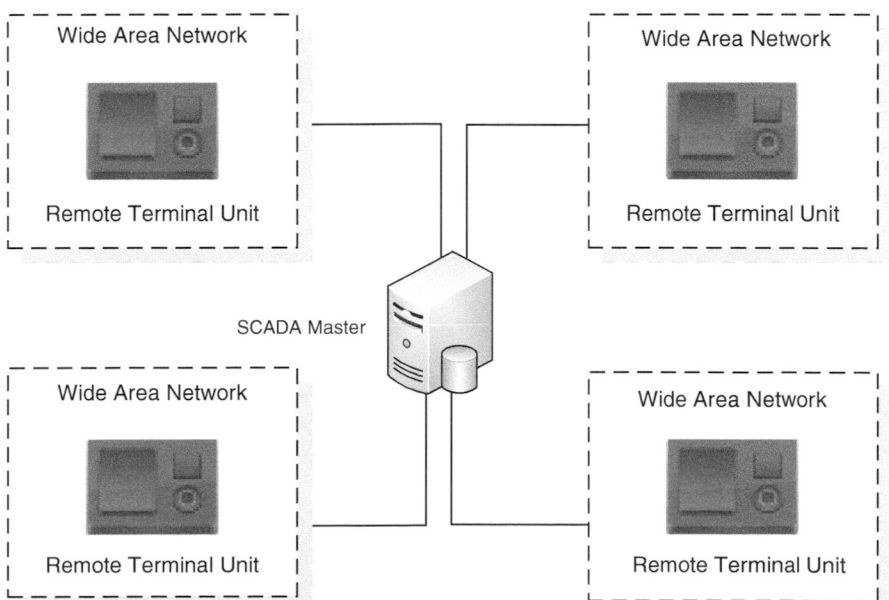

Figure 2.1 First-generation SCADA architecture.

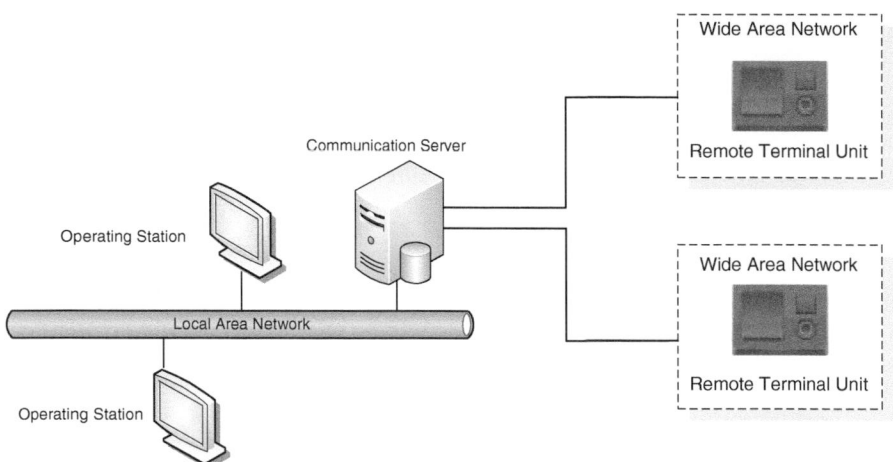

Figure 2.2 Second-generation SCADA architecture.

Networking (LAN) technologies, the SCADA systems of this generation distribute the processing to multiple systems and assigns a specific function for each station. In addition, multiple stations could be connected to an LAN in order to share information with each other in real time. For instance, the communication server can be set up to communicate with field devices such as PLCs and RTUs. Some stations are distributed as MTU, Historian, and HMI servers. The distribution of system functionality across network-connected systems increases processing power, reduces the redundancy, and improves reliability of the system as a whole. In this generation, the system failure is addressed by keeping all stations on the LAN in an online state over the operation time and if one station, say the HMI station, fails, another HMI station will take over.

Networked systems (*Third Generation*). Unlike the second generation, this generation is based on an open system architecture rather than vendor controlled, proprietary solutions. One of the major differences is that the third generation can utilize open standard protocols and products. Consequently, SCADA functionality can be distributed across a WAN and not just a LAN. For instance, most field devices such as PLCs and RTUs can be connected directly to the MTU over an Ethernet connection. This open system architecture allows various products from different vendors to be integrated with each other to build a SCADA system at low cost. In addition, a remote field device can be supervised and controlled from any place and at any time using the Internet. Figure 2.3 shows the architecture of a typical networked SCADA system.

2.1.3 Protocols

There are over 150 protocols utilized by SCADA systems (Igure et al., 2006) but only a small group is widely used. Modbus (IDA, 2004) and DNP3 (Majdalawieh

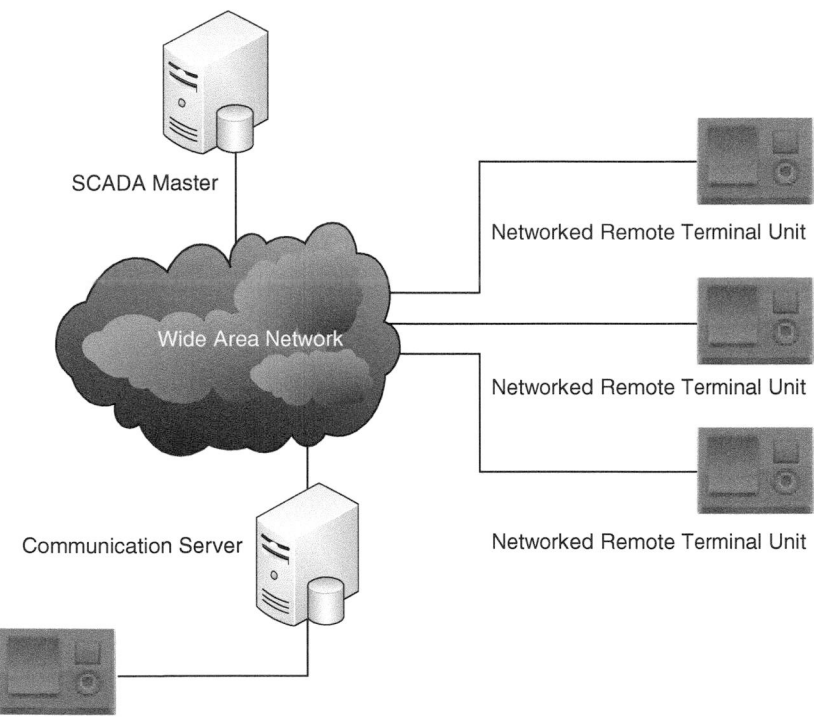

SCADA Master

Networked Remote Terminal Unit

Wide Area Network

Networked Remote Terminal Unit

Communication Server

Networked Remote Terminal Unit

Legacy Remote Terminal Unit

Figure 2.3   Third-generation SCADA architecture.

et al., 2006) are examples of such well-known protocols. The communication proto-col in SCADA is the main weakness regarding security and can be easily attacked from there. Firstly, when the communication protocols were initially proposed for the SCADA network, people were focusing more on their efficiency and effective-ness without considering the potential security issues they might encounter in the future. As the security concerns became critical, security researchers discovered that it was not easy to address this issue. One reason is that the upgrade or replacement of a vital SCADA network in old industrial systems can stop production. Secondly, most of the original SCADA systems were often separate from other corporate networks. Hence, a large number of communication layers and protocols were designed sep-arately, including GE Fanuc, Siemens Sianut, Toshiba, Modbus RTU/ASCII, Allen Bradley DF1/DH, and other vendor protocols.

Modbus is a widely used industrial protocol that works at application level and ensures that data delivery is carried out correctly between devices connected on different kinds of buses or networks. Modbus devices adapted a clientserver approach, where the Modbus slave device represents the server side while the Modbus master device represents the client side of the communication model. Only the master (Client) initiates the communication, while the slave (Server) listens to

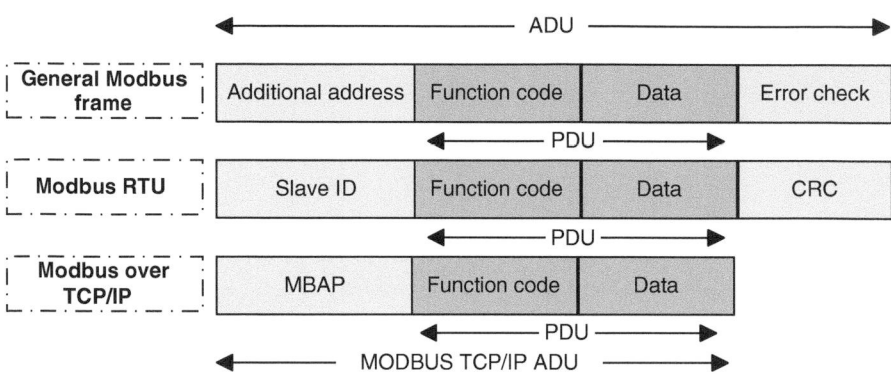

Figure 2.4 The Modbus frame.

the request from the master in order to supply the requested data or execute the requested action. This means Modbus is a request/reply protocol, and has been widely used by millions of automation devices as industry's serial de facto standard communication protocol since 1979. Recently, this protocol has been integrated with TCP/IP and offers a modified version called Modbus/TCP that uses the TCP/IP as transport and network protocols (Modbus Organization, 2020).

Figure 2.4 shows two modes of the Modbus protocol, namely, Modbus RTU and Modbus TCP/IP. The former is the most common implementation and uses binary coding and CRC error-checking. Each message in this mode must be transmitted continuously without inter-character hesitations and is framed by idle (silent) periods. As seen, Modbus PDU includes the Function Code field and Data payload. The server, which listens to any request from the client device, performs actions according to the function codes in the specifications of the protocol. The latter is simply the Modbus RTU protocol with a TCP interface that runs on Ethernet and carries the data of the Modbus message structure between compatible devices and allows them to communicate over a network. As shown in Figure 2.4, a standard Modbus data frame is embedded into a TCP frame without the Modbus checksum because standard Ethernet TCP/IP link layer checksum methods are used to check data integrity.

2.2 INTRUSION DETECTION SYSTEM (IDS)

An IDS is an autonomous hardware or software or a combination of these used to detect threats to SCADA systems from both internal and external attacks, by monitoring and analysing activities on a host computer or a network. A threat can be considered as a malicious activity intended to destroy the security of a SCADA system. Under the threat, the confidentiality, integrity, or availability of the host computers or networks are compromised. In addition, IDS can prevent potential threats to the SCADA system by detecting precursors to an attack, unauthorized access, abnormal operations, etc. According to the location and source of data collected, in traditional IT, IDSs can be categorized into network-based and host-based IDSs (Denning, 1987), and this categorization could be similar even to SCADA systems. However, due to the different nature of SCADA systems in terms of architecture, functionalities, and used devices, SCADA IDSs, within the scope of this book, are categorized based on only the source of data collected: SCADA network-based and SCADA application-based.

2.2.1 SCADA Network-Based

A SCADA network-based IDS (Valdes and Cheung, 2009; Gross et al., 2004; Ning et al., 2002; Linda et al., 2009) captures the data packets that are communicated between devices such as points-to-points in RTU/PLC, between RTU/PLCs and CTUs. The monitoring devices are always located throughout the network. The information in those captured data packets is evaluated to determine whether or not it is a threat. If the packet is suspicious, security team members will be alarmed for further investigation. The advantage of SCADA network-based IDSs is their lower computation cost because only the information in the packet's header is needed for the investigation process, and therefore a SCADA network packet can be scrutinized on-the-fly. Consequently, a large amount of network data can be inspected in a satisfactory manner and within an acceptable time (Linda et al., 2009).

However, when there is high network traffic, a SCADA network-based IDS might experience problems in monitoring all the packets and might miss an attack being launched. The key weakness is that the operational meaning of the monitored SCADA system cannot be inferred from the information provided at the network level such as IP address, TCP port, etc. Therefore, if the payload of the SCADA network packet contains a malicious control message, which is crafted at the application level, the SCADA network-based IDS cannot detect it if it is not violating the specifications of the protocol being used or the communication pattern between SCADA networked devices (Fovino et al., 2010a, 2012; Carcano et al., 2011).

2.2.2 SCADA Application-Based

SCADA applications typically log valuable information about supervised and controlled processes, which are stored in historian servers for maintenance, business, historical, and insight purposes. The SCADA data, which are the measurement and control data generated by sensors and actuators, represent the majority of this information and, in addition, form the operational information for a given SCADA system through which the internal presentation of monitored systems can be inferred (Wenxian and Jiesheng, 2011; Carcano et al., 2011; Fovino et al., 2012; Rrushi et al., 2009b; Zaher et al., 2009). In contrast to the SCADA network-based IDSs that inspect only network level information, a SCADA application-based IDS can inspect high-level data such as SCADA data to detect the presence of unusual behavior. For example, high-level control attacks, which are the most difficult threats to be detected by a SCADA network-based IDS (Wei et al., 2011), can be detected by monitoring the evolution of SCADA data (Rrushi et al., 2009b).

Since the information source of SCADA application-based IDSs can be gathered from different and remote field devices such as PLC and RTU, there are various ways to deploy a SCADA application-based IDS, as follows. (i) It can be deployed in the historian server, as this server is periodically updated by the MTU server which acquires, through field devices such as PLC and RTU, the information and status of the monitored system for each time period. However, this type of deployment raises a security issue, since the real information and statuses in the MTU server can be different from the ones that are sent to the historian server. This could occur when the MTU server is compromised (Jared Verba, 2008). (ii) It can be deployed in an independent server providing that it will not be compromised, and the server from time to time acquires information and statuses from all field devices (Fovino et al., 2010a). Similarly, the large requests from this server each time will increase the network overhead. Consequently, a performance issue may arise. (iii) Each adjacent field device can be connected with a server running SCADA application-based IDS, which are similar to the works in (Alcaraz and Lopez, 2014a, 2014b). However, the key issue is that SCADA data are directly (or indirectly) correlated, and therefore sometimes there is an abnormality in a parameter, not because of itself, but due to a certain value in another parameter (Carcano et al., 2011; Fovino et al., 2012). Therefore, it would be appropriate to assign an individual SCADA application-based IDS for each of the correlated parameters.

2.3   IDS APPROACHES

The concept of IDS is based on the assumption that the behavior of intrusive activities are noticeably distinguishable from the normal ones (Denning, 1987). Many types of SCADA IDSs have been proposed in the literature, and these fall into two broad categories in terms of the detection strategy: *signature-based detection* (Digitalbond, 2013) and *anomaly-based detection* (Linda et al., 2009; Kumar et al., 2007; Valdes and Cheung, 2009; Yang et al., 2006; Ning et al., 2002; Gross et al., 2004).

Signature-based. This approach detects malicious activities in SCADA network traffic or application events by matching the signatures of known attacks that are stored in a specific database. The false positive rate in this type of IDSs is very low and can approach zero. Moreover, the detection time can be fast because it is based only on a matching process in the detection phase. Despite the aforementioned advantages of a signature-based IDS, it will fail to detect an unknown attack whose signature is not known or which does not exist in its database. Therefore, the database must constantly be updated with patterns of new attacks.

SCADA anomaly-based. This approach is based on the assumption that the behavior of intrusive activities mathematically or statistically differs from normal behavior. That is, they are based on advanced mathematical or statistical methods used to detect the abnormal behavior. For example, normal SCADA network traffic can be obtained over a period of "normal" operations, and then a modeling method is applied to build the normal SCADA network profiles. In the detection phase, the deviation degree between the current network flow and the created normal network profiles is calculated. If the deviation degree exceeds the predefined threshold, the current network flow will be flagged as an intrusive activity. The primary advantage of anomaly-based compared to signature-based detection is that novel (unknown) attacks can be detected, although they suffer from a high false positive rate.

A number of factors have a significant impact on the performance of SCADA anomaly-based IDS in distinguishing between the normal and abnormal behavior, including the type of modeling method, the type of building process of the detection models, and the definition of an anomaly threshold. Three learning processes are usually used to build the detection models, namely *supervised*, *semisupervised*, and *unsupervised*. In the supervised learning, anomaly-based IDS requires class labels for both normal and abnormal behavior in order to build normal/abnormal profiles. However, this type of learning is costly and time-expensive when identifying the class labels for a large amount of data. Hence, semisupervised learning is proposed as an alternative, where an anomaly-based IDS builds only normal profiles from the normal data that is collected over a period of "normal" operations. However, the main drawback of this learning is that comprehensive and "purely" normal data is not easy to obtain. This is because the collection of normal data requires that a given system operates under normal conditions for a long time, and intrusive activities may occur during this period of the data collection process. On the another hand, the reliance only on abnormal data for building abnormal profiles is not feasible since the possible abnormal behavior that may occur in the future cannot be known in advance. Alternatively, an anomaly-based IDS uses the unsupervised learning to build normal/abnormal profiles from unlabeled data, where prior knowledge about normal/abnormal data is not known. In fact, it is a cost-efficient method, although it suffers from low efficiency and poor accuracy (Pietro and Mancini, 2008).

CHAPTER 3

SCADA-Based Security Testbed

SCADA systems monitor and supervise our daily infrastructure systems and industrial processes. Hence, the security of such systems cannot be overstated. A number of industrial systems are still using legacy devices, and meanwhile they are directly (or indirectly) connected to the public network (Internet). This is because the sharing of real-time information with business operations has become a necessity for improving efficiency, minimizing costs, and maximizing profits. However this exposes SCADA systems to various types of exploitation. Therefore, it is important to identify common attacks and develop security solutions tailored to SCADA systems. However, to do so, it is impractical to evaluate security solutions on actual live systems. This chapter describes a framework for a SCADA security testbed based on virtualization technology, called SCADAVT, in addition to a server that is used as a surrogate for water distribution systems. Moreover, this chapter presents a case study to demonstrate how the SCADAVT testbed can be effective in monitoring and controlling automated processes, as well as a case study of two well-known malicious attacks to show how such attacks can be launched on supervised processes.

3.1 MOTIVATION

SCADA labs are safe for the evaluation of security solutions, the execution of penetration tests, and the analysis of vulnerabilities and threats. This is because actual SCADA systems are most likely to be affected. Therefore, real SCADA testbeds (e.g., Christiansson and Luiijf (2010) and Sandia National Laboratories (2012), have been designed and set up for evaluation purposes, but these are costly and beyond the reach of most researchers and practionners. Similarly, small real SCADA testbeds (e.g., Fovino et al. (2010b) and Morris et al. (2011) have also been set up, but these are still proprietary- and location-constrained. Hence, a number of model-based SCADA testbeds (e.g., Queiroz et al. (2011), Kush et al. (2010), Davis et al. (2006), Reaves and Morris (2012) and Giani et al. (2008), have been proposed, but they use several modeling tools to build the essential main components of SCADA systems, and the way in which these are linked makes it a complex process to use each testbed. Therefore, such testbeds are unlikely to attract SCADA security experts.

SCADA Security: Machine Learning Concepts for Intrusion Detection and Prevention,
First Edition. Abdulmohsen Almalawi, Zahir Tari, Adil Fahad and Xun Yi.
© 2021 John Wiley & Sons, Inc. Published 2021 by John Wiley & Sons, Inc.

This chapter provides details of SCADAVT, which is a framework for a SCADA testbed that is intended for security experts. Both the essential SCADA components and the controlled environment (e.g. water distribution systems) are properly designed. The former is built on top of the CORE emulator (Ahrenholz, 2010), while the latter is modeled through the use of the dynamic link library (DLL) of EPANET, the well-known modeling tool for simulating water movement and quality behavior within pressurized pipe networks (Lewis, 1999).

The chapter is organized as follows. Section 3.2 presents guidelines for establishing a SCADA security testbed. Section 3.3 describes the guidelines that are implemented to achieve SCADAVT. Section 3.4 presents a scenario application of SCADAVT, and Section 3.5 describes two types of well-known malicious attacks and demonstrates their effects on supervised processes. Finally, conclusions about SCADAVT are provided in Section 3.6.

3.2 GUIDELINES TO BUILDING A SCADA SECURITY TESTBED

The security of SCADA systems has become a topic of interest in recent years, and therefore the number of security solutions has increased. However, the evaluation of their practicality and efficacy is a challenging issue for researchers. This is because the evaluation of actual SCADA systems is not feasible because their availability and performance are most likely to be affected. Moreover, the establishment of actual SCADA security labs is costly, and therefore not affordable for the research community. To address this issue, a SCADA security testbed, consisting of model-based simulation and emulation, can be a realistic alternative to a real-world SCADA system. This section provides a set of guidelines for building a SCADA security testbed.

The development of a simulation of a full SCADA system that realistically mimics a real-world SCADA system consists of two main parts that need to be considered: the controlled environment (e.g., a water distribution system) and the computer-based SCADA components that are responsible for supervising and controlling this environment. The following guidelines are suggested for the development of a simulation that includes these two main parts.

G1. **Select the communication infrastructure.** As previously discussed in Section 2.1.1 (in the Background chapter), the main computer-based SCADA components provide the facility of continuously supervising and controlling the process plant or equipment in industries. Clearly, the communication infrastructure is the first requirement to interconnect these components with each other. To the best of our knowledge, no open-source SCADA network simulator or emulator has been developed from scratch to meet the requirements of SCADA systems. This contrasts with the traditional IT network, where a number of traditional network simulators such as NS2/NS3 (NS3 Maintainers, 2012), OMNET++ (Varga and Hornig, 2008), and QualNet (Developers, 2012), and emulators such as CORE emulator (Ahrenholz, 2010), PlanetLab (Chun et al., 2003), NetBed (Hibler et al., 2008), and NEMAN (Pužar and Plagemann, 2005), are free and publicly available. Since the architectures of networks, hardware, and software of the main SCADA components are computer-based, traditional network simulators and emulators can be adapted and modified to meet SCADA network requirements. Due to the different functionalities of SCADA systems, three features must be available in any candidate traditional network simulator (or emulator): (i) the communication with the external world and the capability of integrating (ii) new protocols and (iii) services (applications).

G2. Develop the main computer-based SCADA components. The main SCADA components are categorized into two levels: high level and field level. The former encompasses SCADA servers that are called Master Terminal Units (MTUs), Human-Machine Interface (HMI), and a historian database; the latter includes field devices such as Remote Terminal Units (RTUs), Programmable Logic Control (PLC), and an Intelligent Electronic Device (IED), which are deployed in the controlled environment to control and collect measurements and control data from sensors and actuators. Since these components are not supported in the traditional network simulators (or emulators), their implementation is important. Hence, an independent simulator for each component needs to be developed. However, several considerations have to be taken into account in the development of each component: the expected functionalities and characteristics; whether the physical location is to be integrated inside the candidate tradition network simulator (or emulator) or placed in the external world; and, finally, whether communication methods will use internal or external components.

G3. Implement SCADA protocols. In contrast to the traditional IT, over 150 protocols are utilized by SCADA systems (Igure et al., 2006). However, only a small group is well known and widely used. Modbus (IDA, 2004) and DNP3 (Majdalawieh et al., 2006) are examples of such protocols. Therefore, the integration of SCADA protocols in the internal world of the traditional candidate network simulator (or emulator) will make it possible to realistically produce SCADA network traffic. In addition, a real SCADA device can communicate with any SCADA component in the simulated (or emulated) network if its communication protocol is supported and the communication between the internal and external world is enabled.

G4. Link the internal SCADA components with the external world. The physical distribution and the functionalities of the computer-based SCADA components must be considered when developing a simulation (or emulation) of a full SCADA system. For instance, field devices such as PLCs and RTUs are computer-based and are distributed in the controlled environment for controlling and collecting measurements and control data from sensors and actuators. The physical location of these devices is supposed to be in the internal-simulated (or emulated) world because they are networked devices (see Figure 3.1). However, it is not feasible to integrate the massive simulated controlled environment (e.g., water distribution system) into the internal world. Therefore, it is practical to implement the controlled environment in the external world and periodically exchange the measurement and control data between it and the field devices. Moreover, the integration of the HMI client, which is one of the main SCADA components, into the internal world may not be practical. This is because this component is responsible for showing the effects of the user's manipulation in a graphical interface for human operators, and therefore the graphical feature may not be supported in the internal world. Finally, the capability of connecting an external SCADA component (whether real or simulated) with the internal world would be a flexible feature.

G5. Simulate a realistic controlled environment. SCADA systems are employed in a number of applications (controlled environments) such as petroleum refining, nuclear power generation, water purification and distribution systems, etc. To fully mimic a SCADA system, this part, which is the controlled environment, needs to be implemented and supervised and controlled by the computer-based SCADA components. For instance, the PowerWorld simulator (Simulator, 2013), which is not free, simulates the power grid with a feature-rich power flow solver, and the process of supervisory control and data acquisition can be performed on this simulator via a TCP-based protocol. Instead, it is time- and cost-efficient to use free and available modeling tools such as the simulations of wind turbine blades (TU Berlin, 2013) or water distribution systems (Lewis, 1999) for simulating the controlled environment.

3.3 SCADAVT DETAILS

Based on the aforementioned guidelines for developing a SCADA security testbed, this section describes a framework for a SCADA security testbed, namely SCA-DAVT. According to these guidelines, an appropriate communication infrastructure is selected for the main computer-based SCADA components that is designed in the second step. In the third step, the well-known and widely used SCADA Modbus/TCP protocol (IDA, 2004) will be integrated into the CORE of the testbed. In the later step, a generic gateway is developed to link the internal world (emulated network) with the external world. Finally, a server that acts as a surrogate for water distribution systems is designed, and this will represent the controlled environment.

3.3.1 The Communication Infrastructure

A thorough investigation is conducted here on a number of traditional network simulators such as NS2/NS3 (NS3 Maintainers, 2012), OMNET++ (Varga and Hornig, 2008), OPNET (OPNET Technologies, 2012), QualNet (Developers, 2012), and emulators such as CORE (Ahrenholz, 2010), PlanetLab (Chun et al., 2003), NetBed (Hibler et al., 2008) and NEMAN (Pužar and Plagemann, 2005). The CORE was chosen as the communication infrastructure for the computer-based SCADA components in SCADAVT. In the following, we will discuss the CORE architecture, followed by the selection features that we considered important when choosing the CORE as the appropriate communication infrastructure for the testbed.

CORE architecture. CORE (Common Open Research Emulator) is a tool for emulating entire traditional networks on one or more machines (Ahrenholz, 2010). Two mechanisms are used in the CORE system: an emulation method (to emulate the routers, PCs, and other hosts) and a simulation method (to simulate the network links between the emulated components). The CORE was derived from the open source Integrated Multi-protocol Emulator Simulator (IMUNES) project from the University of Zagreb (Mikuc et al., 2014). However, the CORE uses FreeBSD and Linux virtualization, as opposed to IMUNES, where only FreeBSD virtualization is used. Figure 3.1 shows the main CORE's components, which are circled by a green dashed-line. To date, the CORE runs only on Linux and FreeBSD systems. In Linux distributions such as Fedora and Ubuntu, the CORE uses Linux network namespaces

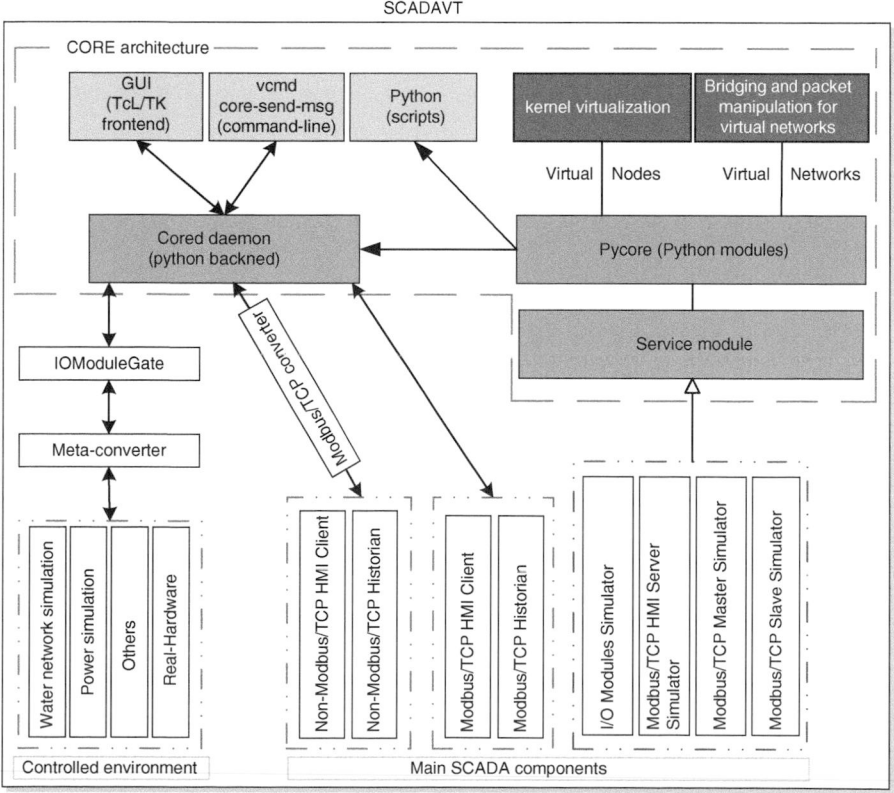

Figure 3.1   SCADAVT Architecture.

as the primary virtualization method to create a namespace that has its own process and private network stack. Afterwards, the CORE combines the created namespaces with Linux Ethernet bridging to form networks.

As shown in Figure 3.1, *Python modules* can be used by a *cored daemon* (back-end) or directly imported by *Python scripts* to manage the emulation sessions. A user can interact with the emulated network only via those components that are colored green. It can also be seen that *Python modules* contain a *Service module*. This module is provided to integrate a new service such as a protocol and application into the CORE. The added service will be a part of the CORE and can be customized for one (or more) virtual nodes. Refer to the CORE documentation for more details (Ahrenholz, 2014).

The selection features. The CORE's features are highlighted here that make it the appropriate candidature network emulator for the proposed SCADA testbed. In addition to the following features, the emphasis is that the CORE has the three necessary features that were previously discussed in the guidelines (see **G1**, Section 3.2).

- `Self-continued networks lab`: All components (e.g., nodes, routers, protocols) that are essential for building an emulated network are available in a friendly development environment that can be used to create any network topology and set up its configuration without the need to write any code.
- `Efficient and scalable`: Tens of nodes can be created using a standard laptop computer. This is because CORE virtualizes some parts of the operating system. That is, only the processes and network stacks are made virtual. Thus, each created node is lighter than a full virtual machine.
- `Distributed with multiple COREs`: This feature increases efficiency and scalability, enabling a number of servers to cooperate in order to run a massive emulated network consisting of thousands of nodes.
- `Highly customizable`: Services such as protocols and applications can be easily customized for each node. That is, the processes or scripts, which run on a node when it is started, can be selected based on the functions required of each node. A number of services are introduced, but if a service that is not available is needed, a new service can be integrated. In fact, this feature can help to integrate the essential SCADA components into the internal world (the emulated network).
- `Connected to live physical networks`: The CORE's emulated network runs in real time. Therefore, it can be connected to live physical networks. This is true for any real network device that has an interface; it will be able to communicate with any CORE node. This can be an advantage when connecting any actual SCADA device with the CORE's nodes for the purpose of security testing, provided that the protocol with which the real SCADA device can communicate is implemented and added to the CORE as a service.
- `Wireless networks are supported`: CORE provides two modes of wireless network emulation: *onoff mode*, where two nodes are linked

with each other based on the distance between them, and *advanced mode*, where the complex effects of communicating wirelessly are emulated. This is performed by the integration of an Extendable Mobile Ad-hoc Network Emulator (EMANE) and CORE (Ahrenholz et al., 2011).

- `It is based on virtualization technology:` The CORE is based on virtualization technology and, because of this, the generated network behavior and data will be similar (as found in Reaves and Morris, 2012) to the ones generated by real systems.

3.3.2 Computer-Based SCADA Components

This section discusses the development and integration of the four essential computer-based SCADA components. Three of them are high-level components: HMI Client, MTU, and HMI servers; the fourth is a field-level component that can be one of the following: PLC, RTU, or IED. In recent times, these field-level devices have nearly the same responsibilities and functionalities. Therefore, PLC is preferred for this testbed, and is sometimes called the slave device. This is because it is only listening for requests from the master. Since, at this stage, only SCADA Modbus/TCP protocol (IDA, 2004) is implemented, all the components are Modbus/TCP-based.

Figure 3.1 illustrates the physical location of each component. A component is considered to be in the internal world only when it is integrated into the CORE and is a part of it; otherwise, it is in the external world. Three SCADA components are integrated into the internal world. For example, *Modbus/TCP Slave Simulator*, *Modbus/TCP Master Simulator*, and *Modbus/TCP HMI Server Simulator* represent a field-level device (e.g. PLC), MTU and HMI servers, respectively. It can also be seen in Figure 3.1 that *I/O Modules Simulator* is integrated into the internal world. This component is designed to act as the real I/O Modules for a field device. In the real world, each field device has I/O Modules that are used to interface sensors and actuators in the controlled environment. Therefore, *I/O Modules Simulator* can be used with any simulated field device to synchronize the measurement and control data of the controlled environment that is located in the external world, with the simulated field device in the internal world.

The CORE architecture has a *Pycore* component that contains a number of Python modules. The service module is one of these, whereby a new service (e.g., application, process, or protocol) can be integrated into the CORE using Python scripts. Therefore, we implement all the components of the internal world using Python Programming Language, integrated as services.

On the another hand, two SCADA components (e.g. *Modbus/TCP Historian* and *Modbus/TCP HMI Client*) are proposed to be in the external world (see the justifications in Section 3.2, **G4**). In this testbed, similar to the HMI client component, we use a publicly available Modbus/TCP HMI client (Software Development Team, 2014). As shown in Figure 3.1, SCADA Modbus/TCP-based components in the external world can directly connect with others in the internal world. This is because all SCADA components in the internal world support the SCADA Modbus/TCP protocol.

Modbus/TCP Simulators of Master/Slave. Here the implementation of each component is discussed. For clarity, *Modbus/TCP Master Simulator* and *Modbus/TCP Slave Simulator* represent the MTU server and field device (e.g., PLC), respectively. The modern SCADA systems adapted a master–slave model that is similar to the client–server approach, where the role of the slave model is to listen to any request from the master model. The latter sends control messages to a number of slaves to which a required slave responds according to the control instructions received. Since SCADAVT in this stage supports only the Modbus/TCP protocol, these components are Modbus/TCP-based (see Figure 3.1).

To implement these two SCADA components, Python-based master/slave simulators are designed. The publicly-available library of Modbus/TCP protocol (Team, 2014), which is Python-based, is imported in each simulator so that any Modbus/TCP-based component, whether real or virtual, can communicate with it. The integration of the library of Modbus/TCP protocol is discussed in the following step, which concerns the implementation of SCADA protocols.

Similar to any actual Modbus-field device (e.g. PLC or RTU), *Modbus/TCP Slave simulator* will require mapping its registers prior to being run for the purpose of controlling and monitoring. Therefore, a registers map procedure is provided for mapping the registers as follows:

```
pro_1 = ['ProcessID1','C',1,'i'];
pro_2 = ['ProcessID2','C',2,'i'];
...
...
pro_9 = ['ProcessID9','C',10,'o'];
Registers.add([pro_1,pro_2,..,pro_9]);
```

Pro 1 is a Python list variable that contains the tag of a supervised process parameter and its block type and position in the RAM in its associated slave simulator and parameter type (e.g, input/output). Four symbols, namely H, C, D, and A, are used to represent the following register types, HOLDING, COILS, DISCRETE, and ANALOG registers, respectively. It can be seen from the above that the types of registers are COIL. The last line is the function that adds the mapped registers. All IDs of processes have to be unique. This is because the gateway reads and writes measurement data from and to the registers in each *Modbus/TCP Slave Simulator* through the ID process (see Section 3.3.4).

Execute function, which is provided by Modbus/TCP library (Team, 2014), is used in *Modbus/TCP Master Simulator* for pushing and pulling the measurement and control data to and from *Modbus/TCP Slave Simulator*.

Modbus/TCP Simulator of HMI Server. The HMI server is an intermediate component between the MTU and HMI client where the HMI client sends the user's manipulation to the HMI server in order to be read and executed by the MTU. In the opposite direction, the MTU sends to the HMI server the collected data from a field device after the user's manipulation so that the HMI client can request it in order to show the effects of the user's manipulation in a graphical interface for human operators. As can be seen in Figure 3.1, the HMI client is an external component in the SCADAVT testbed because the HMI client with a graphical interface cannot be supported in the CORE. Therefore, a Python-based simulator of the Modbus/TCP HMI server is implemented to run two independent instances: the first instance listens to the request from the MTU via the internal IP of the virtual node in the emulated environment; the second instance listens to the request from the HMI client via the *backchannel*, which is assigned to each emulated node. Therefore, the HMI client can connect with the Modbus/TCP HMI server in two ways: via a *backchannel* or directly if the HMI client has an independent physical interface and also supports the Modbus/TCP protocol.

I/O Modules Simulator. Modbus/TCP slave simulator, which will be running in the virtual node, is required to monitor and control the simulated supervised process such as the simulations of power generation and water distribution systems that are outside the emulated environment. Therefore, the Python-based *IOModules* simulator is implemented and integrated into the CORE, where it acts as a server that receives input data from the external environment and sends output data when requested. This is carried out through the *backchannel* for each virtual node using a simple and intuitive custom TCP-based protocol called *IOModules*, which will be elaborated on in Section 3.3.4.

3.3.3 SCADA Protocols's Implementation

Over 150 protocols are utilized by SCADA systems (Igure et al., 2006) and, therefore, it is challenging to implement them all. However, there are several well-known and widely used protocols such as Modbus (IDA, 2004), DNP3 (Majdalawieh et al., 2006), and Zigbee (Cunha et al., 2007). At this stage, only the Modbus protocol is supported. This protocol used to work only on Modicon programmable controllers. However, it has become widely used in recent SCADA control devices. Modbus devices adapted a client–server approach, where the Modbus slave device represents the server side while the Modbus master device represents the client side of the communication model. Only the master (Client) initiates the communication, while the slave (Server) listens to the request from the master in order to supply the requested data or execute the requested action.

Thanks to open and free software, the Python-based library of the Modbus/TCP protocol, which is available under GNU (General Public Licence) (Team, 2014), is used. This library is distributed as Python modules, and it needs to be installed on the platform hosting the CORE. To use this library from the CORE, a Python-based script that imports this library needs to be created and added as a new service to the CORE. For example, the Modbus/TCP SCADA components such as *Modbus/TCP Master*

TABLE 3.1   A Screenshot of the Configuration of IOModuleGate

First configuration file	Second configuration file
[PLC1]	[t2_level]
ip:172.16.0.1	controller : PLC1
port:9161	paraType : i
[PLC2]	[t2_level]
ip:172.16.0.2	controller : PLC1
port:9161	paraType : o

Simulator and *Modbus/TCP Slave Simulator*, which are discussed in Section 3.3.2, import this library.

3.3.4   Linking Internal/External World Components

As previously discussed in Section 3.2, **G4**, the physical location of several computer-based SCADA components such as field devices are proposed to be integrated into the internal world since they are networked devices. However, they are required to control and collect measurement and control data from sensors and actuators that are distributed in the controlled environment in the external world. Thus, the two need to be linked. This section shows how a generic gateway, Python-based class, called *IOModuleGate*, is implemented. *IOModules* protocol, which will be discussed later, is implemented to periodically exchange the measurement data between *Modbus/TCP slave simulator* and the respective supervised process parameters through *I/O Modules Simulator* running in each virtual node. Two configuration files are invoked to *IOModuleGate*, where each file is formatted as shown in Table 3.1. For example, in the first file, PLC1 represents the ID for *Modbus/TCP Slave Simulator* whose *backchannel* IP and port are 172.16.0.1 and 9161, respectively. However, in the second file, the process parameters P1 and P2 are supervised by this *Modbus/TCP slave simulator* (PLC1). The parameter type is indicated by either "o" or "i" (e.g., input/output). For pulling and pushing the measurement data to and from the emulated environment, two public writing and reading methods are provided by *IOModuleGate*. These methods take and return a Python dictionary variable, which is a key-value pair. The key is the identity ID of the process parameter (e.g., P1) and its I/O data, where each process parameter in a supervised process must have a unique ID.

The IOModules Protocol   A simple custom TCP-based protocol is implemented in *I/O modules simulator* and *IOModuleGate* to exchange measurement and control data. Figure 3.2 shows four fields that comprise the message structure defined as follows. (1) *TransactionNo*: a unique number for each reading and writing operation. Both reading and writing operations have independent sequential numbering and initially start with one. In the response message, this field contains the same

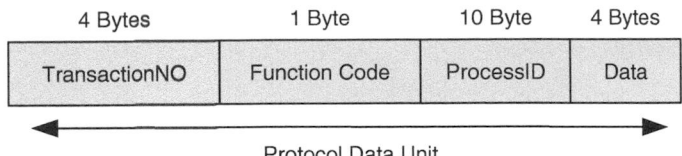

Figure 3.2 IOModules protocol message structure.

number of request messages to indicate that output data is available and correctly read. However, if it contains zero, this indicates that the output data is not ready to be read, and therefore it needs to wait a while before requesting again. The amount of waiting time can be specified in the initialisation time of the IOModuleGate class. (2) *Function Code*: this takes three values, 1, 2, and 0, to indicate reading, writing, and termination operations respectively. (3) *ProcessID*: each process parameter in a supervised process must have a unique ID. (4) *Data*: it contains the process parameter's value. This field is set to zero in the request message of the reading operation.

3.3.5 Simulation of a Controlled Environment

Real-world SCADA systems are intended to supervise and control industrial processes and utilities (e.g. petroleum refining, water distribution systems, etc.). In Section 3.2, **G5**, the simulation of a controlled environment is identified as the second important part of the development of a full SCADA solution. This section introduces a server that simulates water distribution systems using a dynamic link library (DLL) of the well-known modeling tool, which is called EPANET, for simulating water movement and quality behavior within pressurized pipe networks (Lewis, 1999). This server is designed using Visual Basic 6 language. Three items are required: (i) the description file, which can be designed and exported by the visual interface of the EPANET tool, which describes the topology and properties of all components for the simulated water distribution system; (ii) the port number on which the server is listening; and (iii) the time interval to recompute new simulated data.

The server is provided with a custom TCP-based protocol to manipulate the simulated data using a SCADA system. For instance, a client can acquire and control pump status in the simulation through this protocol. Figure 3.3 shows the message structure of this protocol. (1) *ProcessID* contains the ID of the process parameter, which needs to be manipulated. (2) *Function code* defines only two operation types:

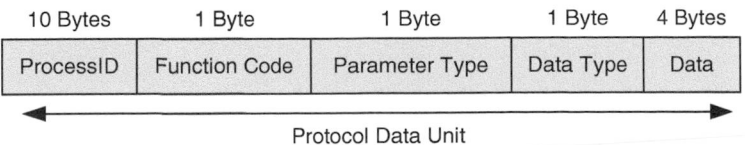

Figure 3.3 The protocol message structure of the WaterSystem Server.

acquisition and control. (3) *Parameter Type* defines only two component types, *Node* and *link*. (4) *Data Type* specifies the data type that needs to be manipulated. For example, the link component such as a pump has a number of data types that can be manipulated (e.g., speed, pumping energy, and status). (5) *Data* contains the process parameter's value. This field is set to zero in the request message for the reading operation.

3.4   SCADAVT APPLICATION

This section presents the details of a real-life application of SCADAVT, where four steps are performed. Firstly, the important steps to set up SCADAVT are presented in bullet points. Secondly, the various parts (e.g., the properties, equipment, expected services) of the water distribution system, which comprise the controlled environment in this scenario, are defined. In the remaining steps (e.g., the characteristics, configurations, topology) of the SCADA system responsible for supervising and controlling the aforementioned controlled environment are described.

3.4.1   The SCADAVT Setup

To set up SCADAVT, a number of dependencies are required to be installed as follows:

- CORE is the core part of SCADAVT. For installation details, please refer to Ahrenholz (2010).
- The Modbus library is a library provided by a third party and is publicly available (Team, 2014).
- Python interpreter is a prerequisite for the CORE, the Modbus library, and our integration scripts.
- The hpin3 utility is a security assignment tool that is used to assemble/analyze TCP/IP packets (Sanfilippo, 2012). This tool supports many protocols such as TCP, UDP, and ICMP. Moreover, this tool can be used to launch a number of attacks such as Denial of Service and spoofing attacks.
- The integration scripts are our Python-based scripts developed in order to integrate the essential SCADA components into the CORE as services. To automatically add these services, a user needs to move these scripts to the *myservices* directory, which is found in the CORE directory path prior to starting up the CORE.

3.4.2   The Water Distribution System Setup

As shown in Figure 3.4, a Water Distribution System (WDS) is designed for a small town using the graphical interface of the EPANET tool. The water network is divided

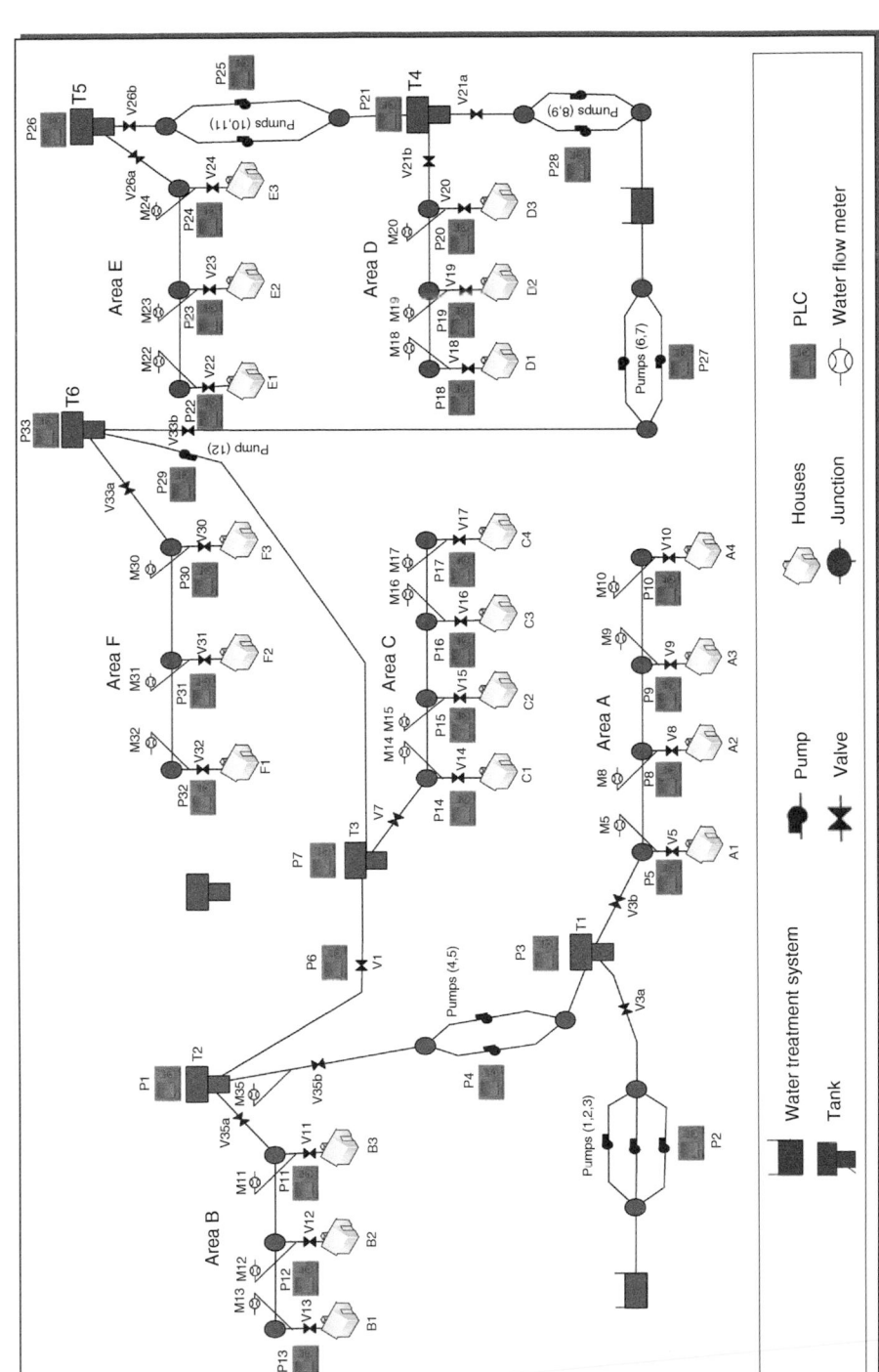

Figure 3.4 The simulation of a water distribution system.

into six areas, namely A, B, C, D, E, and F. Each area has an elevated tank to supply the area with water at a satisfactory pressure level. There are two water treatment systems to supply water to all areas. The first supplies areas A, B, and C; the second supplies areas D, E, and F. In addition, area C can be supplied from the second system through area F, which is subjected to some considerations of water demands in both areas C and F. The supplied water is pumped out by three pumps from the first water treatment system into $Tank_1$. The water is also delivered to $Tank_2$ by two pumps. $Tank_3$ is supplied through gravity because the elevation of $Tank_2$ is higher than $Tank_3$. $Tank_1$ is twice the size of $Tank_2$ and $Tank_3$ because it is the main water source for areas B and C. Similarly, four pumps are used to pump out the supplied water from the second water treatment system. The first two pumps deliver the water to $Tank_4$, while the other two pumps deliver the water to $Tank_6$. $Tank_5$ is supplied with water from $Tank_4$ using two pumps. Each area consists of a number of sub-areas, which are represented by a house symbol. A number of houses (households) are located in each sub-area, and the number of people who live in each sub-area is randomly selected, provided that the expected demands of the people do not exceed the capacity that the system can deliver. In fact, people are not always at home for the whole time; therefore, every hour, we randomly assign a small fraction of the number of people in each sub-area as people who are away. This fraction does not exceed 10%. The distribution of people throughout the areas is shown in Table 3.2.

The water consumption in the water network model is one of the factors that reflects the behavior of simulated data. Therefore, a realistic model of water consumption behavior is required in order to obtain more realistic simulated data. Therefore, the consumption module is fed, in the EPANET model, with a specific model based on Melbourne Water (2009) (i.e., the 2009–2010 Melbourne water consumption). Table 3.3 illustrates Melbournians' average water usage per day over one week. To roughly estimate the average water usage by one person per day, the local mean μ and standard deviation σ are computed for each sequential n week. Then, a random number is generated, which represents the consumption of water per person per day, from the normal distribution with mean μ and standard deviation σ. Let X be a vector of average water usage per day over n weeks, where $X = \{x_1, x_2, \dots, x_n\}$.

TABLE 3.2   The distribution of people throughout the areas

Area	Sub-area	People	Area	Sub-area	People	Area	Sub-area	People
A	A1	4500	C	C1	3600	E	E1	3560
	A2	5200		C2	4999		E2	4500
	A3	5000		C3	2560		E3	3690
	A4	3500		C4	3256			
B	B1	6000	D	D1	6800	F	F1	3600
	B2	3500		D2	4600		F2	4300
	B3	6500		D3	3600		F3	3600

TABLE 3.3   Melburnians' Average Water Usage per Day over Week

Date	Avg-C	Date	Avg-C	Date	Avg-C	Date	Avg-C
9/01/2009	143	29/05/2009	142	23/10/2009	146	12/03/2010	133
16/01/2009	172	5/06/2009	134	30/10/2009	157	19/03/2010	155
23/01/2009	188	12/06/2009	129	6/11/2009	150	26/03/2010	152
30/01/2009	207	19/06/2009	136	13/11/2009	204	2/04/2010	144
6/02/2009	241	26/06/2009	136	20/11/2009	200	9/04/2010	141
13/02/2009	202	3/07/2009	130	27/11/2009	166	16/04/2010	138
20/02/2009	195	17/07/2009	136	4/12/2009	145	23/04/2010	154
27/02/2009	186	24/07/2009	143	11/12/2009	153	30/04/2010	135
6/03/2009	181	31/07/2009	141	18/12/2009	165	7/05/2010	139
13/03/2009	149	7/08/2009	139	25/12/2009	165	14/05/2010	136
20/03/2009	134	14/08/2009	138	1/01/2010	138	21/05/2010	136
27/03/2009	152	21/08/2009	141	8/01/2010	138	28/05/2010	135
3/04/2009	155	28/08/2009	138	15/01/2010	187	4/06/2010	133
10/04/2009	143	4/09/2009	141	22/01/2010	150	11/06/2010	129
17/04/2009	135	11/09/2009	139	29/01/2010	157	18/06/2010	126
24/04/2009	151	18/09/2009	150	5/02/2010	189	25/06/2010	131
1/05/2009	133	25/09/2009	137	12/02/2010	170	2/07/2010	127
8/05/2009	138	2/10/2009	132	19/02/2010	141	9/07/2010	129
15/05/2009	135	9/10/2009	138	26/02/2010	174	16/07/2010	135
22/05/2009	140	16/10/2009	144	5/03/2010	164	23/07/2010	132

Avg-C: Average water usage per day over week.

The mean μ and the standard deviation σ of X are obtained with the following equations:

$$\mu = \frac{1}{n} \sum_{i=1}^{n} X(i) \tag{3.1}$$

$$\sigma = \frac{1}{n} \sqrt{\sum_{i=1}^{n} (X(i) - \mu)^2} \tag{3.2}$$

For instance, let the current week in the simulation be $23 - 01 - 2009$ and n be five weeks, which are the current week and two weeks before and after. It is assumed here that five weeks can represent seasonal variations in generating daily consumption. The average consumption per day for a person over this week and two weeks before $(09 - 01 - 2009, 16 - 01 - 2009)$ and after $(30 - 01 - 2009, 06 - 02 - 2009)$ are $X = \{143, 172, 188, 207, 241\}$. To calculate μ and σ of X Equations 5.8 and 3.2 are used, respectively. The consumption per day for a person in this week is randomly generated from a normal distribution with mean 190.2 and standard deviation 36.83.

The hourly water consumption is determined by dividing the period of 24 hours into four classes as follows:

1. $\{0, 1, 2, 3, 4, 5\}$: this class consumes 10% and the consumption for each hour in this class is equal, that is, each hourly consumption is 0.0166%.

2. $\{6, 7, 8, 9, 10, 11\}$: this class consumes 35%, the consumption for each hour in this class is different, and is as follows: 0.10%, 0.10%,0.05%, 0.033%, 0.033%,0.033%, respectively.

3. $\{12, 13, 14, 15, 16, 17\}$: this class consumes 20% and the consumption for each hour in this group is equal to 0.033% for each.

4. $\{18, 19, 20, 21, 22, 23\}$: this class consumes 35%, the consumption for each hour in this class is different, and is as follows: 0.10%, 0.10%,0.05%, 0.033%, 0.033%,0.033%, respectively.

These classes are based on our assumptions about routine daily usage, which would be typical for each time period of the day.

3.4.3   SCADA System Setup for WDS

This section shows how the SCADA devices are used to monitor and control the previously discussed water distribution system. This process is carried out by dragging and dropping the components of the CORE such as virtual node, link, and router. The integrated SCADA components (e.g., Modbus/TCP slave simulator) are automatically added to the services that can be assigned to any virtual node with one click. Figure 3.5 shows the SCADA network topology for this scenario. Thirty-three PLCs are deployed throughout fourteen field areas (see Table 3.4) and each PLC is connected with a field router to communicate with the Master Terminal Unit (MTU) over the Internet, which is represented (in this scenario) by a number of connected routers.

Each PLC is assigned to specific functions in this scenario. Table 3.5 shows end devices (sensors or actuators) that are supervised and controlled by each PLC. All these PLCs are managed by the MTU, which is represented by a virtual node that required the service of the Modbus/TCP Master Simulator. Through the MTU, in this scenario, a number of functions can be remotely performed either manually (via the HMI) or automatically (via the MTU). In the following, Table 3.6 summarizes all these instructions that the MTU can perform via each PLC.

Algorithm 1 is implemented to maintain sufficient water in both $Tank_2$ and $Tank_3$. This problem is illustrated in Figure 3.6, where the water level of $Tank_2$ reached the critical level seven times, during which area C was not efficiently supplied with water. This is because the flow valve V_1 has a fixed setting, which is 1300 liters per minute (LPM) in this example, and the water flow from $Tank_2$ to $Tank_3$ is constant, even though the water level in $Tank_2$ is low. This problem is addressed by considering the following parameters: the water level in $Tank_2$ and $Tank_3$, the current water demands in areas C and B, and the water flow pumped in to $Tank_2$. These parameters are used by the MTU to intelligently adjust the flow valve V_1. Figure 3.7 shows the water level of $Tank_2$ and $Tank_3$ after applying this algorithm. Thanks to SCADA systems, there is an increase in the performance of daily services using less equipment.

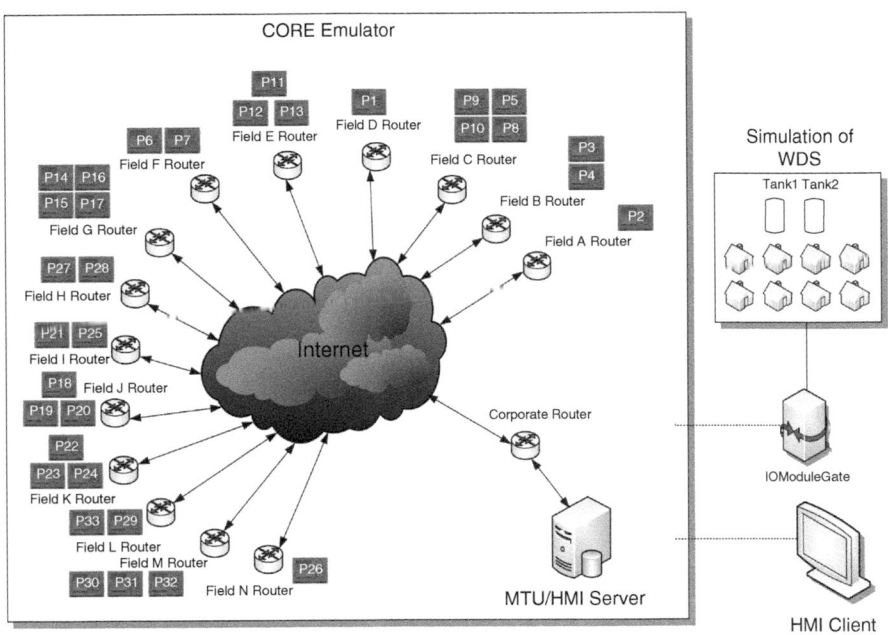

Figure 3.5 SCADA network topology for controlling the scenario of the water distribution network.

TABLE 3.4 **The Deployment of PLCs over Field Areas**

Field	Controllers	Field	Controllers
A	PLC2	H	PLC27, PLC28
B	PLC3, PLC4	I	PLC21, PLC25
C	PLC5, PLC8, PLC9, PLC10	J	PLC18, PLC19, PLC20
D	PLC1	K	PLC22, PLC23, PLC24
E	PLC11, PLC12, PLC13	L	PLC33, PLC29
F	PLC6, PLC7	M	PLC30, PLC31, PLC32
G	PLC14, PLC15, PLC16, PLC17	N	PLC26

3.4.4 Configuration Steps

To control and monitor processes using SCADAVT, a number of configuration steps are required similar to any actual SCADA system. Both I/O modules and Modbus/TCP slave simulator services are enabled for each virtual node in order to represent a PLC. Then, the IP address and port are assigned. Since the PLC is used to read and control end devices (sensors and actuators), the map of the registers of the PLC with its associated end devices needs to be configured. Tables 3.7, 3.8, 3.9, and 3.10 later in Appendix 3.7.1 have mapping of Modbus

TABLE 3.5   Field Devices and Their Respective Supervised Devices

Field device	Supervised devices
PLC1	Water level sensor of tank (T2) and actuators of valves (V1a and V1b)
PLC2	Actuators of pumps (P1, P2, P3)
PLC3	Water level sensor of tank (T1) and actuators of valves (V3a and V3b)
PLC4	Actuators of pumps (P4, P5)
PLC5	Actuator of valve (V5) and flow meter (M5)
PLC6	Actuator of valve (V1)
PLC7	Water level sensor of tank (T3)
PLC8	Actuator of valve (V8) and flow meter (M8)
PLC9	Actuator of valve (V9) and flow meter (M9)
PLC10	Actuator of valve (V10) and flow meter (M10)
PLC11	Actuator of valve (V11) and flow meter (M11)
PLC12	Actuator of valve (V12) and flow meter (M12)
PLC13	Actuator of valve (V13) and flow meter (M13)
PLC14	Actuator of valve (V14) and flow meter (M14)
PLC15	Actuator of valve (V15) and flow meter (M15)
PLC16	Actuator of valve (V16) and flow meter (M16)
PLC17	Actuator of valve (V17) and flow meter (M17)
PLC18	Actuator of valve (V18) and flow meter (M18)
PLC19	Actuator of valve (V19) and flow meter (M19)
PLC20	Actuator of valve (V20) and flow meter (M20)
PLC21	Water level sensor of tank (T4) and actuators of valves (V21a and V21b)
PLC22	Actuator of valve (V22) and flow meter (M22)
PLC23	Actuator of valve (V23) and flow meter (M23)
PLC24	Actuator of valve (V24) and flow meter (M24)
PLC25	Actuators of pumps (P10, P11)
PLC26	Water level sensor of tank (T5) and actuators of valves (V26a and V26b)
PLC27	Actuators of pumps (P6, P7)
PLC28	Actuators of pumps (P8, P9)
PLC29	Actuator of pump (P12)
PLC30	Actuator of valve (V30) and flow meter (M30)
PLC31	Actuator of valve (V31) and flow meter (M31)
PLC32	Actuator of valve (V32) and flow meter (M32)
PLC33	Water level sensor of tank (T6) and actuator of valves (V33)

registers for all PLCS used in the implemented scenario. For example, Table 3.7 shows that the first resister of ANALOG registers in the field device PLC1 is assigned to the $t2\_level$ (the water level sensor of $tank_2$). Therefore, through the *IOModuleGate* the data of the $t2\_level$ can be sent to the first ANALOG register in the PLC1. In this case, the MTU device can read this register using the *execute* function provided by the Modbus library, Team (2014).

To exchange the management and control data between the emulated SCADA system and the supervised environment, *IOModuleGate* class is extended. However, two configuration files need to be given when the *IOModuleGate* class is initialized.

TABLE 3.6 The Control and Monitoring Instructions that MTU Performs Through Each PLC

Field device	Instructions
PLC1	Reads the water level in tank T2 and controls the valves V35a, V35b
PLC2	Reads status, speed and energy of the pumps P1, P2, P2 and controls their statuses and speeds
PLC3	Reads the water level in tank T1 and controls the valves V3a, V3b
PLC4	Reads status, speed and energy of the pumps P1, P2 and controls their statuses and speeds
PLC5	Reads water flow and presuure at sub-area A1 using flow meter M5 and the presuure sensor attached in V5, and controls the presuure using V5
PLC6	Intelligently adjusts the flow valve V1 , which is between tank T2 and tank T3, with an appropriate setting. This is done using Algorithm
PLC7	Reads the water level in tank T3 and controls the valves V7
PLC8	Reads water flow and presuure at sub-area A2 using flow meter M8 and the presuure sensor attached in V8, and controls the presuure using V8
PLC9	Reads water flow and presuure at sub-area A3 using flow meter M9 and the presuure sensor attached in V9, and controls the presuure using V9
PLC10	Reads water flow and presuure at sub-area A4 using flow meter M10 and the presuure sensor attached in V10, and controls the presuure using V10
PLC11	Reads water flow and presuure at sub-area B3 using flow meter M11 and the presuure sensor attached in V11, and controls the presuure using V11
PLC12	Reads water flow and presuure at sub-area B2 using flow meter M12 and the presuure sensor attached in V12, and controls the presuure using V12
PLC13	Reads water flow and presuure at sub-area B1 using flow meter M13 and the presuure sensor attached in V13, and controls the presuure using V13
PLC14	Reads water flow and presuure at sub-area C1 using flow meter M14 and the presuure sensor attached in V14, and controls the presuure using V14
PLC15	Reads water flow and presuure at sub-area C2 using flow meter M15 and the presuure sensor attached in V15, and controls the presuure using V15
PLC16	Reads water flow and presuure at sub-area C3 using flow meter M16 and the presuure sensor attached in V16, and controls the presuure using V16
PLC17	Reads water flow and presuure at sub-area C4 using flow meter M17 and the presuure sensor attached in V17, and controls the presuure using V17
PLC18	Reads water flow and presuure at sub-area D1 using flow meter M18 and the presuure sensor attached in V18, and controls the presuure using V18
PLC19	Reads water flow and presuure at sub-area D2 using flow meter M19 and the presuure sensor attached in V19, and controls the presuure using V19
PLC20	Reads water flow and presuure at sub-area D3 using flow meter M20 and the presuure sensor attached in V20, and controls the presuure using V20
PLC21	Reads the water level in tank T4 and controls the valves V21a, V21b
PLC22	Reads water flow and presuure at sub-area E1 using flow meter M22 and the presuure sensor attached in V22, and controls the presuure using V22
PLC23	Reads water flow and presuure at sub-area E2 using flow meter M23 and the presuure sensor attached in V23, and controls the presuure using V23

(Continued)

TABLE 3.6   (Continued)

Field device	Instructions
PLC24	Reads water flow and presuure at sub-area E3 using flow meter M24 and the presuure sensor attached in V24, and controls the pressure using V24
PLC25	Reads status, speed and energy of the pumps P10, P11 and controls their statuses and speeds
PLC26	Reads the water level in tank T5 and controls the valves V26a, V26b
PLC27	Reads status, speed and energy of the pumps P6, P7 and controls their statuses and speeds
PLC28	Reads status, speed and energy of the pumps P8, P9 and controls their statuses and speeds
PLC29	Reads status, speed and energy of the pump P12 and controls its status and speed
PLC30	Reads water flow and presuure at sub-area F3 using flow meter M30 and the presuure sensor attached in V30, and controls the presuure using V30
PLC31	Reads water flow and presuure at sub-area F2 using flow meter M31 and the presuure sensor attached in V31, and controls the presuure using V31
PLC32	Reads water flow and presuure at sub-area F1 using flow meter M32 and the presuure sensor attached in V32, and controls the presuure using V32
PLC33	Reads the water level in tank T6 and controls the valves V33a, V33b

Note: status of a pump either OFF or ON.

Algorithm 1   A smart control algorithm controlling water flow from $Tank_2$ to $Tank_3$.

---

1 $b \Leftarrow$ *Water demand in area B*
2 $c \Leftarrow$ *Water demand in area C*
3 $t_2 \Leftarrow$ *Water level in tank$_2$*
4 $t_3 \Leftarrow$ *Water level in tank$_3$*
5 $f \Leftarrow$ *Water flow to tank$_2$*
6 **if** $t_2 > t_3$ **then**
7 $\quad$ $flow = b + (f - c)$
8 $\quad$ *Adjust V_1 to flow*
9 **else**
10 $\quad$ $flow = b - c$
11 $\quad$ *Adjust V_1 to flow*

The first file contains a list of *backchannel* IP addresses and ports for all PLCs. Notably, the IP addresses are not the IPs that are used within the emulated control system, but are the IPs that are automatically assigned by the CORE emulator for each emulated device as a *backchannel* to allow communication with the external world. Therefore, this feature is used by the I/O modules service to exchange

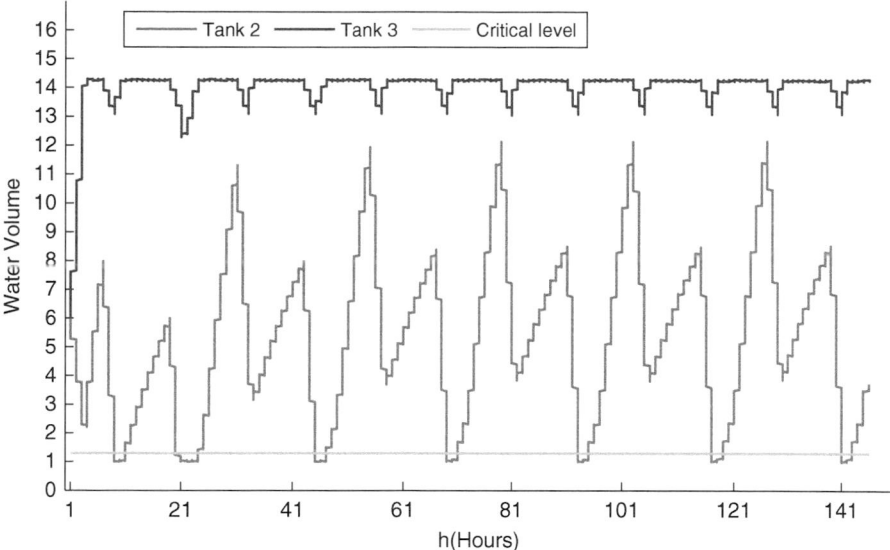

Figure 3.6   The water levels over a period of time for *Tank₂* and *Tank₃* without control system.

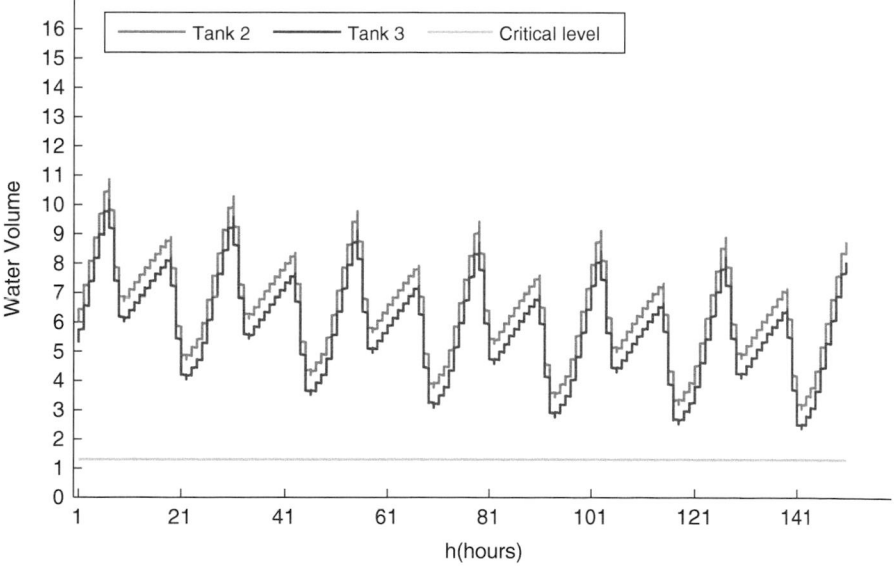

Figure 3.7   The water levels over a period of time for *Tank₂* and *Tank₃* with control system.

the measurement data between supervised processes and the emulated SCADA system through the *IOModuleGate*. The second file contains a list of process parameters and the IDs of their associated field devices. Tables 3.11, 3.12, 3.13, and 3.14 later in Appendix 3.7.2 show the configuration files for the implemented scenario. For example, in the first file, PLC1 represents the ID for a field device whose *backchannel* IP and port are 172.16.0.1 and 9161, respectively. In the second file, the process parameter *t2_level* (the water level sensor of *tank$_2$*) is supervised by the field device PLC1 and the data type *dataType* of *t2_level* is input, where input indicates that the end device can be read only by a PLC, while output indicates that a PLC can change the data of the end device.

To realistically mimic SCADA systems, the HMI client needs to be integrated to allow an operator to monitor and manually control the process. Therefore, we developed an HMI server that supports the Modbus protocol as an integrated service in the proposed SCADAVT. For example, in Figure 3.5 this service was added to the MTU device, which is set up to listen on the *backchannel* of the MTU. Therefore, any HMI client who supports the Modbus protocol can directly connect to the HMI server. In this scenario, the publicly available HMI client Software Development Team (2014) is used.

To start exchanging measurement data between the control system and the *IOModuleGate*, writing/reading methods provided by the IOModuleGate are used. These methods take and return a Python dictionary variable, which is a key-value pair. The key is the identity ID of the process parameter (e.g *t2_level*) and its I/O data. In this scenario, a connection to the server of the WDS simulation is established, and this requires the simulated data through its TCP-custom-protocol (see Section 3.3.5). Afterwards, we construct a Python dictionary variable with input data and push it by writing methods to respective field devices (PLCs). Similarly, we pull the output data from the field devices using the reading method, and then push them to their respective process parameters. The period of time for pulling and pushing the data to and from a control system and supervised infrastructure is determined by the implementer of the IOModuleGate class.

As is clear from the detailed discussion above, the functioning of the emulation requires several different configuration steps and this in turn requires specific knowledge of SCADA systems. Therefore, the user of SCADAVT who performs the simulation has to be a person who is well-versed in the specifics of SCADA systems.

3.5   ATTACK SCENARIOS

The evaluation of SCADAVT is carried out through two common attack scenarios: a *denial of service* and an *integrity attack*. This section also shows how these malicious attacks can affect WDS's performance.

3.5.1   Denial of Service (DoS) Attacks

In this type of attack, attackers launch flood attacks against a target to prevent it from receiving a legitimate request. As previously discussed, the MTU periodically adjusts the flow valve V_1, which is controlled by PLC_6, using Algorithm ??. In fact, if the MTU cannot establish a connection with PLC_6 to send a control message, the flow valve V_1 will not be properly adjusted. Hence, the water volume in *Tank$_2$* and *Tank$_3$* will not be balanced, so the critical level may be reached. In this scenario, we demonstrate a Distributed Denial of Service attack (DDoS) where 10 virtual nodes are attached to the Internet, which is represented in this scenario by a set of routers linked with each other. This is easily done with a few clicks. The open source hping3 utility (Sanfilippo, 2012) is used to launch flood attacks on the field device *PLC$_6$*. Three times *PLC$_6$* was flooded with TCP SYN packets. The first attack started at time= 15 h and ended at 20 h. The second attack started at time= 55 h and ended at 57 h. The last attack started at time= 100 h and ended at 105 h. During these attack times, the MTU sometimes failed to establish a connection with *PLC$_6$* and sometimes it took a long time to successfully connect with it. Figure 3.8 shows the unsuccessful and successful connections between MTU and *PLC$_6$*. It can be seen that the unsuccessful connections and the connection establishing time at the period time of DDoS are significantly different from normal behavior. That is, the MTU failed a number of times and waited a long time to establish connection compared to attack-free time. Hence, because the MTU failed to intelligently adjust the flow valve V_1, the water volumes of both *Tank$_2$* and *Tank$_3$* have been affected. Figure 3.9 clearly shows that the water volumes of *Tank$_2$* and *Tank$_3$* fell below the critical level twice and once, respectively. Consequently, areas C and B were not sufficiently supplied with water twice and once, respectively.

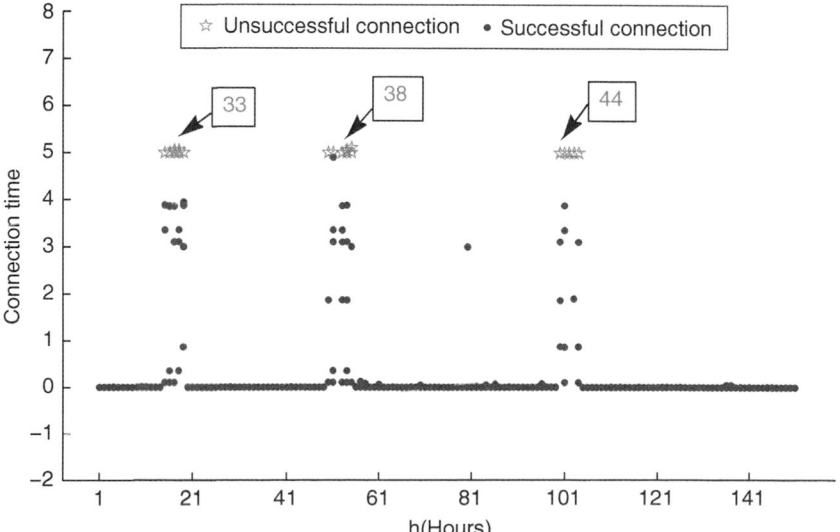

Figure 3.8 The unsuccessful and successful connections and their elapsed times between MTU and *PLC$_6$*.

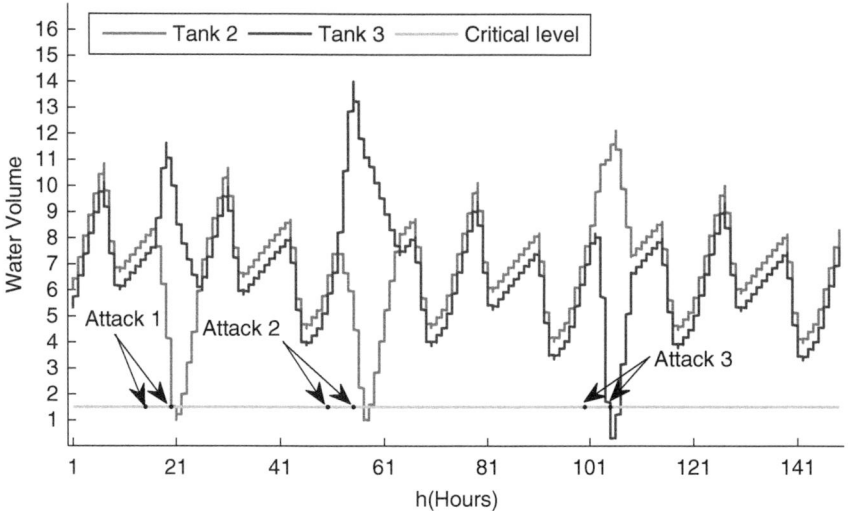

Figure 3.9   The effect of DDoS, which targets PLC_6, on the water volume of $Tank_2$ and $Tank_3$.

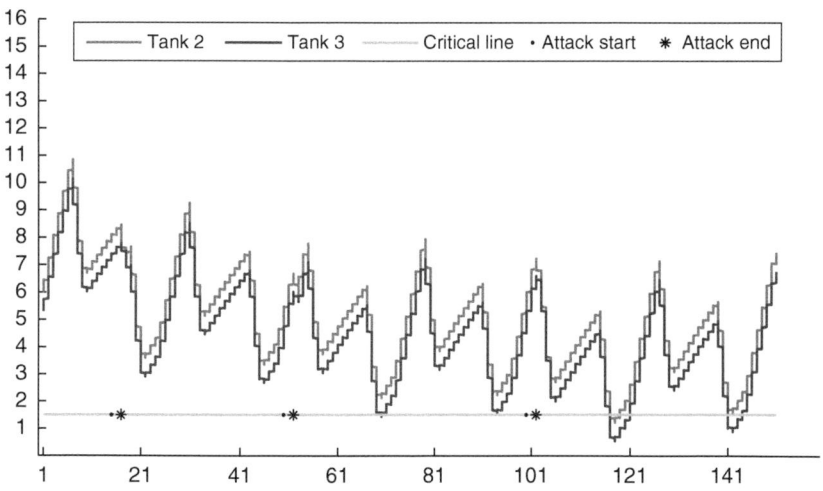

Figure 3.10   The effect of an integrity attack, which targets PLC_4, on the water volume of $Tank_2$ and $Tank_3$.

3.5.2   Integrity Attacks

This type of attack occurs as a result of the manipulation of command messages; it is termed a *high-level control attack* (Queiroz et al., 2011; Wei et al., 2011; Giani et al., 2013). To launch such an attack, an attacker requires prior knowledge of the target system. This can be obtained by using the specifications, or by a correlation analysis of the network traffic of that system. Taking over the control center and sending undesired control messages, or intercepting (e.g., man-in-middle attack) the command messages between the control center and field devices, are a common means of launching such attacks. In fact, such an attack is difficult to detect because the false message is still legitimate in terms of the Modbus/TCP protocol specifications. To demonstrate this type of attack, the control message between the MTU and field device PLC_4 is intercepted and modified, which controls the operation of $Pump_4$ and $Pump_5$. To perform this attack, a proxy between these devices is established. As previously discussed, the MTU sends a control message to PLC_4 to turn its asso ciated pumps ON/OFF. The intercepted control message sent to PLC_4 three times is modified, The starting and ending times of each integrity attack are depicted in Figure 3.10. In each attack, the intercepted control message with a control data is modified, whereby $Pump_4$ and $Pump_5$ are turned off. Figure 3.10 shows the water volumes of $Tank_2$ and $Tank_3$ after the integrity attack, and it can be seen that the critical level was reached several times, and that the effect of attacks has not occurred at the same times as the attacks. This depends on the functionality being attacked in the target system.

3.6 CONCLUSION

This chapter presented SCADAVT, a framework for a SCADA security testbed based on virtualisation technology. SCADAVT is a novel solution to create a full SCADA system based on emulation and simulation methods. In addition, it realistically mimics the real SCADA testbed, and also has a feature that allows an actual SCADA device to be connected for the purpose of realistic evaluation. Furthermore, a server, which acts as a surrogate for water distribution systems, is introduced. A scenario application of SCADAVT is presented to show how the testbed can easily be used to create a realistic simulation of a full SCADA system. DDoS and integrity attacks have been described to illustrate how malicious attacks can be launched on supervised processes.

3.7 APPENDIX FOR THIS CHAPTER

3.7.1 Modbus Registers Mapping

TABLE 3.7 **Mapping of Modbus Registers to the Process Parameters of the Implemented Scenario**

Field device	Registers mapping

PLC1	```var1 = ['t2_level','A',0,'i'];```
	```var2 = ['m35_flow','A',1,'i'];```
	```var3 = ['v35a_settings_i','A',2,'i'];```
	```var4 = ['v35a_settings_o','A',3,'o'];```
	```var5 = ['v35b_settings_i','A',4,'i'];```
	```var6 = ['v35b_settings_o','A',5,'o'];```
	```regMap.add([var1,var2,var3,var4,var5,var6]);```

PLC2	```var1 = ['p1_status_i','D',0,'i'];```
	```var2 = ['p1_status_o','D',1,'o'];```
	```var3 = ['p1_speed_i','A',0,'i'];```
	```var4 = ['p1_speed_o','A',1,'o'];```
	```var5 = ['p1_ energy _i','A',2,'i'];```
	```var6 = ['p2_status_i','D',2,'i'];```
	```var7 = ['p2_status_o','D',3,'o'];```
	```var8 = ['p2_speed_i','A',3,'i'];```
	```var9 = ['p2_speed_o','A',4,'o'];```
	```var10 = ['p2_ energy _i','A',5,'i'];```
	```var11 = ['p3_status_i','D',3,'i'];```
	```var12 = ['p3_status_o','D',4,'o'];```
	```var13 = ['p3_speed_i','A',6,'i'];```
	```var14 = ['p3_speed_o','A',7,'o'];```
	```var15 = ['p3_ energy _i','A',8,'i'];```
	```regMap.add([var1,var2,var3,...,var15]);```

PLC3	```var1 = ['t1_level','A',0,'i'];```
	```var2 = ['v3a_settings_i','A',1,'i'];```
	```var3 = ['v3a_settings_o','A',2,'o'];```
	```var4 = ['v3b_settings_i','A',3,'i'];```
	```var5 = ['v3b_settings_o','A',4,'o'];```
	```regMap.add([var1,var2,var3,var4,var5]);```

TABLE 3.7 (Continued)

Field device	Registers mapping
| PLC4 | ```
var1 = ['p4_status_i','D',0,'i'];
var2 = ['p4_status_o','D',1,'o'];
var3 = ['p4_speed_i','A',0,'i'];
var4 = ['p4_speed_o','A',1,'o'];
var5 = ['p4_ energy _i','A',2,'i'];
var6 = ['p5_status_i','D',2,'i'];
var7 = ['p5_status_o','D',3,'o'];
var8 = ['p5_speed_i','A',3,'i'];
var9 = ['p5_speed_o','A',4,'o'];
var10 = ['p5_ energy _i','A',5,'i'];
regMap.add([var1,var2,var3,...,var10]);
``` |
| PLC5 | ```
var1 = ['m5_flow','A',0,'i'];
var2 = ['v5_ pressure','A',1,'i'];
var3 = ['v5_settings_i','A',2,'i'];
var4 = ['v5_settings_o','A',3,'o'];
regMap.add([var1,var2,var3,var4]);
``` |
| PLC6 | ```
var1 = ['m5_flow','A',0,'i'];
var2 = ['v5_ pressure','A',1,'i'];
var3 = ['v5_settings_i','A',2,'i'];
var4 = ['v5_settings_o','A',3,'o'];
regMap.add([var1,var2,var3,var4]);
``` |
| PLC6 | ```
var1 = ['m5_flow','A',0,'i'];
var2 = ['v5_ pressure','A',1,'i'];
var3 = ['v5_settings_i','A',2,'i'];
var4 = ['v5_settings_o','A',3,'o'];
regMap.add([var1,var2,var3,var4]);
``` |
| PLC6 | ```
var1 = ['v1_settings_i','A',0,'i'];
var2 = ['v1_settings_o','A',1,'o'];
regMap.add([var1,var2]);
``` |
| PLC7 | ```
var1 = ['t3_level','A',0,'i'];
var2 = ['v7_settings_i','A',1,'i'];
var3 = ['v7_settings_o','A',2,'o'];
regMap.add([var1,var2,var3]);
``` |
| PLC8 | ```
var1 = ['m8_flow','A',0,'i'];
var2 = ['v8_ pressure','A',1,'i'];
var3 = ['v8_settings_i','A',2,'i'];
var4 = ['v8_settings_o','A',3,'o'];
regMap.add([var1,var2,var3,var4]);
``` |
```

TABLE 3.8   Mapping of Modbus Registers to the Process Parameters of the Implemented Scenario

Field device	Registers mapping
| PLC9 | ```
var1 = ['m9_flow','A',0,'i'];
var2 = ['v9_ pressure','A',1,'i'];
var3 = ['v9_settings_i','A',2,'i'];
var4 = ['v9_settings_o','A',3,'o'];
regMap.add([var1,var2,var3,var4]);
``` |
| PLC10 | ```
var1 = ['m10_flow','A',0,'i'];
var2 = ['v10_ pressure','A',1,'i'];
var3 = ['v10_settings_i','A',2,'i'];
var4 = ['v10_settings_o','A',3,'o'];
regMap.add([var1,var2,var3,var4]);
``` |
| PLC11 | ```
var1 = ['m11_flow','A',0,'i'];
var2 = ['v11_ pressure','A',1,'i'];
var3 = ['v11_settings_i','A',2,'i'];
var4 = ['v11_settings_o','A',3,'o'];
regMap.add([var1,var2,var3,var4]);
``` |
| PLC12 | ```
var1 = ['m12_flow','A',0,'i'];
var2 = ['v12_ pressure','A',1,'i'];
var3 = ['v12_settings_i','A',2,'i'];
var4 = ['v12_settings_o','A',3,'o'];
regMap.add([var1,var2,var3,var4]);
``` |
| PLC13 | ```
var1 = ['m13_flow','A',0,'i'];
var2 = ['v13_ pressure','A',1,'i'];
var3 = ['v13_settings_i','A',2,'i'];
var4 = ['v13_settings_o','A',3,'o'];
regMap.add([var1,var2,var3,var4]);
``` |
| PLC14 | ```
var1 = ['m14_flow','A',0,'i'];
var2 = ['v14_ pressure','A',1,'i'];
var3 = ['v14_settings_i','A',2,'i'];
var4 = ['v14_settings_o','A',3,'o'];
regMap.add([var1,var2,var3,var4]);
``` |
| PLC15 | ```
var1 = ['m15_flow','A',0,'i'];
var2 = ['v15_ pressure','A',1,'i'];
var3 = ['v15_settings_i','A',2,'i'];
var4 = ['v15_settings_o','A',3,'o'];
regMap.add([var1,var2,var3,var4]);
``` |
```

TABLE 3.8   (Continued)

Field device	Registers mapping

PLC16	var1 = ['m16_flow','A',0,'i'];
	var2 = ['v16_ pressure','A',1,'i'];
	var3 = ['v16_settings_i','A',2,'i'];
	var4 = ['v16_settings_o','A',3,'o'];
	regMap.add([var1,var2,var3,var4]);

PLC17	var1 = ['m17_flow','A',0,'i'];
	var2 = ['v17_ pressure','A',1,'i'];
	var3 = ['v17_settings_i','A',2,'i'];
	var4 = ['v17_settings_o','A',3,'o'];
	regMap.add([var1,var2,var3,var4]);

PLC18	var1 = ['m18_flow','A',0,'i'];
	var2 = ['v18_ pressure','A',1,'i'];
	var3 = ['v18_settings_i','A',2,'i'];
	var4 = ['v18_settings_o','A',3,'o'];
	regMap.add([var1,var2,var3,var4]);

TABLE 3.9   **Mapping of Modbus Registers to the Process Parameters of the Implemented Scenario**

Field device	Registers mapping

PLC19	var1 = ['m19_flow','A',0,'i'];
	var2 = ['v19_ pressure','A',1,'i'];
	var3 = ['v19_settings_i','A',2,'i'];
	var4 = ['v19_settings_o','A',3,'o'];
	regMap.add([var1,var2,var3,var4]);

PLC20	var1 = ['m20_flow','A',0,'i'];
	var2 = ['v20_ pressure','A',1,'i'];
	var3 = ['v20_settings_i','A',2,'i'];
	var4 = ['v20_settings_o','A',3,'o'];
	regMap.add([var1,var2,var3,var4]);

PLC21	var1 = ['t4_level','A',0,'i'];
	var2 = ['v21a_settings_i','A',2,'i'];
	var3 = ['v21a_settings_o','A',3,'o'];
	var4 = ['v21b_settings_i','A',4,'i'];
	var5 = ['v21b_settings_o','A',5,'o'];
	regMap.add([var1,var2,var3,var4,var5]);

(Continued)

TABLE 3.9   (Continued)

Field device	Registers mapping

PLC22	```varl = ['m22_flow','A',0,'i'];```
	```var2 = ['v22_ pressure','A',1,'i'];```
	```var3 = ['v22_settings_i','A',2,'i'];```
	```var4 = ['v22_settings_o','A',3,'o'];```
	```regMap.add([varl,var2,var3,var4]);```

PLC23	```varl = ['m23_flow','A',0,'i'];```
	```var2 = ['v23_ pressure','A',1,'i'];```
	```var3 = ['v23_settings_i','A',2,'i'];```
	```var4 = ['v23_settings_o','A',3,'o'];```
	```regMap.add([varl,var2,var3,var4]);```

PLC24	```varl = ['m24_flow','A',0,'i'];```
	```var2 = ['v24_ pressure','A',1,'i'];```
	```var3 = ['v24_settings_i','A',2,'i'];```
	```var4 = ['v24_settings_o','A',3,'o'];```
	```regMap.add([varl,var2,var3,var4]);```

PLC25	```varl = ['p10_status_i','D',0,'i'];```
	```var2 = ['p10_status_o','D',1,'o'];```
	```var3 = ['p10_speed_i','A',0,'i'];```
	```var4 = ['p10_speed_o','A',1,'o'];```
	```var5 = ['p10_ energy _i','A',2,'i'];```
	```var6 = ['p11_status_i','D',2,'i'];```
	```var7 = ['p11_status_o','D',3,'o'];```
	```var8 = ['p11_speed_i','A',3,'i'];```
	```var9 = ['p11_speed_o','A',4,'o'];```
	```var10 = ['p11_ energy _i','A',5,'i'];```
	```regMap.add([varl,var2,var3,...,var10]);```

PLC26	```varl = ['t5_level','A',0,'i'];```
	```var2 = ['v26a_settings_i','A',2,'i'];```
	```var3 = ['v26a_settings_o','A',3,'o'];```
	```var4 = ['v26b_settings_i','A',4,'i'];```
	```var5 = ['v26b_settings_o','A',5,'o'];```
	```regMap.add([varl,var2,var3,var4,var5]);```

TABLE 3.10   Mapping of Modbus Registers to the Process Parameters of the Implemented Scenario

Field device	Registers mapping

PLC27	var1 = ['p6_status_i','D',0,'i'];
	var2 = ['p6_status_o','D',1,'o'];
	var3 = ['p6_speed_i','A',0,'i'];
	var4 = ['p6_speed_o','A',1,'o'];
	var5 = ['p6_ energy _i','A',2,'i'];
	var6 = ['P7_status_i','D',2,'i'];
	var7 = ['p7_status_o','D',3,'o'];
	var8 = ['p7_speed_i','A',3,'i'];
	var9 = ['p7_speed_o','A',4,'o'];
	var10 = ['p7_ energy _i','A',5,'i'];
	regMap.add([var1,var2,var3,...,var10]);

PLC28	var1 = ['p8_status_i','D',0,'i'];
	var2 = ['p8_status_o','D',1,'o'];
	var3 = ['p8_speed_i','A',0,'i'];
	var4 = ['p8_speed_o','A',1,'o'];
	var5 = ['p8_ energy _i','A',2,'i'];
	var6 = ['p9_status_i','D',2,'i'];
	var7 = ['p9_status_o','D',3,'o'];
	var8 = ['p9_speed_i','A',3,'i'];
	var9 = ['p9_speed_o','A',4,'o'];
	var10 = ['p9_ energy _i','A',5,'i'];
	regMap.add([var1,var2,var3,...,var10]);

PLC29	var1 = ['p12_status_i','D',0,'i'];
	var2 = ['p12_status_o','D',1,'o'];
	var3 = ['p12_speed_i','A',0,'i'];
	var4 = ['p12_speed_o','A',1,'o'];
	var5 = ['p12_ energy _i','A',2,'i'];
	regMap.add([var1,var2,var3,var4,var5]);

(Continued)

TABLE 3.10   (Continued)

Field device	Registers mapping

PLC30	```varl = ['m30_flow','A',0,'i'];```
	```var2 = ['v30_ pressure','A',1,'i'];```
	```var3 = ['v30_settings_i','A',2,'i'];```
	```var4 = ['v30_settings_o','A',3,'o'];```
	```regMap.add([varl,var2,var3,var4]);```

PLC31	```varl = ['m31_flow','A',0,'i'];```
	```var2 = ['v31_ pressure','A',1,'i'];```
	```var3 = ['v31_settings_i','A',2,'i'];```
	```var4 = ['v31_settings_o','A',3,'o'];```
	```regMap.add([varl,var2,var3,var4]);```

PLC32	```varl = ['m32_flow','A',0,'i'];```
	```var2 = ['v32_ pressure','A',1,'i'];```
	```var3 = ['v32_settings_i','A',2,'i'];```
	```var4 = ['v32_settings_o','A',3,'o'];```
	```regMap.add([varl,var2,var3,var4]);```

PLC33	```varl = ['t6_level','A',0,'i'];```
	```var2 = ['v33a_settings_i','A',2,'i'];```
	```var3 = ['v33a_settings_o','A',3,'o'];```
	```var4 = ['v33b_settings_i','A',4,'i'];```
	```var5 = ['v33b_settings_o','A',5,'o'];```
	```regMap.add([varl,var2,var3,var4,var5]);```

3.7.2 The Configuration of IOModuleGate

TABLE 3.11   Mapping of Modbus Registers to the Process Parameters of the Implemented Scenario

First configuration file

[PLC1]	[PLC12]	[PLC23]
ip:172.16.0.1	ip:172.16.0.12	ip:172.16.0.23
port:9161	port:9161	port:9161
[PLC2]	[PLC13]	[PLC24]
ip:172.16.0.2	ip:172.16.0.13	ip:172.16.0.24
port:9161	port:9161	port:9161
[PLC3]	[PLC14]	[PLC25]
ip:172.16.0.3	ip:172.16.0.14	ip:172.16.0.25
port:9161	port:9161	port:9161
[PLC4]	[PLC15]	[PLC26]
ip:172.16.0.4	ip:172.16.0.14	ip:172.16.0.26
port:9161	port:9161	port:9161
[PLC5]	[PLC16]	[PLC27]
ip:172.16.0.5	ip:172.16.0.16	ip:172.16.0.27
port:9161	port:9161	port:9161
[PLC6]	[PLC17]	[PLC28]
ip:172.16.0.6	ip:172.16.0.17	ip:172.16.0.28
port:9161	port:9161	port:9161
[PLC7]	[PLC18]	[PLC29]
ip:172.16.0.7	ip:172.16.0.18	ip:172.16.0.29
port:9161	port:9161	port:9161
[PLC8]	[PLC19]	[PLC30]
ip:172.16.0.8	ip:172.16.0.19	ip:172.16.0.30
port:9161	port:9161	port:9161
[PLC9]	[PLC20]	[PLC31]
ip:172.16.0.9	ip:172.16.0.20	ip:172.16.0.31
port:9161	port:9161	port:9161
[PLC10]	[PLC21]	[PLC32]
ip:172.16.0.10	ip:172.16.0.21	ip:172.16.0.32
port:9161	port:9161	port:9161
[PLC11]	[PLC22]	[PLC33]
ip:172.16.0.11	ip:172.16.0.22	ip:172.16.0.33
port:9161	port:9161	port:9161

TABLE 3.12 Mapping of Modbus Registers to the Process Parameters of the Implemented Scenario

Second configuration file

```
[t2_level]                [p3_speed_i]              [m5_flow]                 [v10_pressure]
    controller : PLC1          controller : PLC2         controller : PLC5         controller : PLC10
    dataType : i               dataType : i              dataType : i              dataType : i
[m35_flow]                [p3_speed_o]              [v5_pressure]             [v10_settings_i]
    controller : PLC1          controller : PLC2         controller : PLC5         controller : PLC10
    dataType : i               dataType : o              dataType : i              dataType : i
[v35a_settings_i]         [p3_energy_i]             [v5_settings_i]           [v10_settings_o]
    controller : PLC1          controller : PLC2         controller : PLC5         controller : PLC10
    dataType : i               dataType : i              dataType : i              dataType : o
[v35a_settings_o]         [t1_level]                [v5_settings_o]           [m11_flow]
    controller : PLC1          controller : PLC1         controller : PLC5         controller : PLC11
    dataType : o               dataType : o              dataType : o              dataType : i
[v35b_settings_i]         [v3a_settings_i]          [v1_settings_i]           [v11_pressure]
    controller : PLC1          controller : PLC3         controller : PLC6         controller : PLC11
    dataType : i               dataType : i              dataType : i              dataType : i
[v35b_settings_o]         [v3a_settings_o]          [v1_settings_o]           [v11_settings_i]
    controller : PLC1          controller : PLC3         controller : PLC6         controller : PLC11
    dataType : o               dataType : o              dataType : o              dataType : i
[p1_status_i]             [v3b_settings_i]          [t3_level]                [v11_settings_o]
    controller : PLC2          controller : PLC3         controller : PLC7         controller : PLC11
    dataType : i               dataType : i              dataType : i              dataType : o
[p1_status_o]             [v3b_settings_o]          [v7_settings_i]           [m12_flow]
    controller : PLC2          controller : PLC3         controller : PLC7         controller : PLC12
    dataType : o               dataType : o              dataType : i              dataType : i
[p1_speed_i]              [p4_status_i]             [v7_settings_o]           [v12_pressure]
    controller : PLC2          controller : PLC4         controller : PLC7         controller : PLC12
    dataType : i               dataType : i              dataType : o              dataType : i
```

```
[p1_speed_o]
    controller : PLC2
    dataType : o
[p1_ energy _i]
    controller : PLC2
    dataType : i
[p2_status_i]
    controller : PLC2
    dataType : i
[p2_status_o]
    controller : PLC2
    dataType : o
[p2_speed_i]
    controller : PLC2
    dataType : i
[p2_speed_o]
    controller : PLC2
    dataType : o
[p2_ energy _i]
    controller : PLC2
    dataType : i
[p3_status_i]
    controller : PLC2
    dataType : i
[p3_status_o]
    controller : PLC2
    dataType : o

[p4_status_o]
    controller : PLC4
    dataType : o
[p4_speed_i]
    controller : PLC4
    dataType : i
[p4_speed_o]
    controller : PLC4
    dataType : o
[p4_ energy _i]
    controller : PLC4
    dataType : i
[p5_status_i]
    controller : PLC4
    dataType : i
[p5_status_o]
    controller : PLC4
    dataType : o
[p5_speed_i]
    controller : PLC4
    dataType : i
[p5_speed_o]
    controller : PLC4
    dataType : o
[p5_ energy _i]
    controller : PLC4
    dataType : i

[m8_flow]
    controller : PLC8
    dataType : i
[8_ pressure]
    controller : PLC8
    dataType : i
[v8_settings_i]
    controller : PLC8
    dataType : i
[v8_settings_o]
    controller : PLC8
    dataType : o
[m9_flow]
    controller : PLC9
    dataType : i
[v9_ pressure]
    controller : PLC9
    dataType : i
[v9_settings_i]
    controller : PLC9
    dataType : i
[v9_settings_o]
    controller : PLC9
    dataType : o
[m10_flow]
    controller : PLC10
    dataType : i

[v12_settings_i]
    controller : PLC12
    dataType : i
[v12_settings_o]
    controller : PLC12
    dataType : o
[m13_flow]
    controller : PLC13
    dataType : i
[v13_ pressure]
    controller : PLC13
    dataType : i
[v13_settings_i]
    controller : PLC13
    dataType : i
[v13_settings_o]
    controller : PLC13
    dataType : o
[m14_flow]
    controller : PLC14
    dataType : i
[v14_ pressure]
    controller : PLC14
    dataType : i
[v14_settings_i]
    controller : PLC14
    dataType : i
```

TABLE 3.13 Mapping of Modbus Registers to the Process Parameters of the Implemented Scenario

Second configuration file

```
[v14_settings_o]              [v19_pressure]                [v23_settings_i]              [v26a_settings_o]
    controller : PLC14            controller : PLC19            controller : PLC23            controller : PLC26
    dataType : o                  dataType : i                  dataType : i                  dataType : o
[m15_flow]                    [v19_settings_i]              [v23_settings_o]              [v26b_settings_i]
    controller : PLC15            controller : PLC19            controller : PLC23            controller : PLC26
    dataType : i                  dataType : i                  dataType : o                  dataType : i
[v15_pressure]                [v19_settings_o]              [m24_flow]                    [v26b_settings_o]
    controller : PLC15            controller : PLC19            controller : PLC24            controller : PLC26
    dataType : i                  dataType : o                  dataType : i                  dataType : o
[v15_settings_o]              [m20_flow]                    [v24_pressure]                [p6_status_i]
    controller : PLC15            controller : PLC20            controller : PLC24            controller : PLC27
    dataType : o                  dataType : i                  dataType : i                  dataType : i
[v15_settings_i]              [v20_pressure]                [v24_settings_i]              [p6_status_o]
    controller : PLC15            controller : PLC20            controller : PLC24            controller : PLC27
    dataType : i                  dataType : i                  dataType : i                  dataType : o
[m16_flow]                    [v20_settings_o]              [v24_settings_o]              [p6_speed_i]
    controller : PLC16            controller : PLC20            controller : PLC24            controller : PLC27
    dataType : i                  dataType : o                  dataType : o                  dataType : i
[v16_pressure]                [v20_settings_i]              [p10_status_i]                [p6_speed_o]
    controller : PLC16            controller : PLC20            controller : PLC25            controller : PLC27
    dataType : i                  dataType : i                  dataType : i                  dataType : o
[v16_settings_i]              [t4_level]                    [p10_status_o]                [p6_energy_i]
    controller : PLC16            controller : PLC21            controller : PLC25            controller : PLC27
    dataType : i                  dataType : i                  dataType : o                  dataType : i
[v16_settings_o]              [v21a_settings_i]             [p10_speed_i]                 [p7_status_i]
    controller : PLC16            controller : PLC21            controller : PLC25            controller : PLC27
    dataType : o                  dataType : i                  dataType : i                  dataType : i
```

```
[m17_flow]
    controller : PLC17
    dataType : i
[v17_pressure]
    controller : PLC17
    dataType : i
[v17_settings_i]
    controller : PLC17
    dataType : i
[v17_settings_o]
    controller : PLC17
    dataType : o
[m18_flow]
    controller : PLC18
    dataType : i
[v18_pressure]
    controller : PLC18
    dataType : i
[v18_settings_i]
    controller : PLC18
    dataType : i
[v18_settings_o]
    controller : PLC18
    dataType : o
[m19_flow]
    controller : PLC19
    dataType : i

[v21a_settings_o]
    controller : PLC21
    dataType : o
[v21b_settings_i]
    controller : PLC21
    dataType : i
[v21b_settings_o]
    controller : PLC21
    dataType : o
[m22_flow]
    controller : PLC22
    dataType : i
[v22_pressure]
    controller : PLC22
    dataType : i
[v22_settings_i]
    controller : PLC22
    dataType : i
[v22_settings_o]
    controller : PLC22
    dataType : o
[m23_flow]
    controller : PLC23
    dataType : i
[v23_pressure]
    controller : PLC23
    dataType : i

[p10_speed_o]
    controller : PLC25
    dataType : o
[p10_energy_i]
    controller : PLC25
    dataType : i
[p11_status_i]
    controller : PLC25
    dataType : i
[p11_status_o]
    controller : PLC25
    dataType : o
[p11_speed_i]
    controller : PLC25
    dataType : i
[p11_speed_o]
    controller : PLC25
    dataType : o
[p11_energy_i]
    controller : PLC25
    dataType : i
[t5_level]
    controller : PLC26
    dataType : i
[v26a_settings_i]
    controller : PLC26
    dataType : i

[p7_status_o]
    controller : PLC27
    dataType : o
[p7_speed_i]
    controller : PLC27
    dataType : i
[p7_speed_o]
    controller : PLC27
    dataType : o
[p7_energy_i]
    controller : PLC27
    dataType : i
[p8_status_i]
    controller : PLC28
    dataType : i
[p8_status_o]
    controller : PLC28
    dataType : o
[p8_speed_i]
    controller : PLC28
    dataType : i
[p8_speed_o]
    controller : PLC28
    dataType : o
[p8_energy_i]
    controller : PLC28
    dataType : i
```

TABLE 3.14 Mapping of Modbus Registers to the Process Parameters of the Implemented Scenario

Second configuration file

```
[p9_status_i]            [p12_speed_i]            [m31_flow]               [v32_settings_o]
    controller : PLC28       controller : PLC29       controller : PLC31       controller : PLC32
    dataType : i             dataType : i             dataType : i             dataType : o
[p9_status_o]            [p12_speed_o]            [v31_pressure]           [t6_level]
    controller : PLC28       controller : PLC29       controller : PLC31       controller : PLC33
    dataType : o             dataType : o             dataType : i             dataType : i
[p9_speed_i]             [p12_ energy _i]         [v31_settings_i]         [v33a_settings_i]
    controller : PLC28       controller : PLC29       controller : PLC31       controller : PLC33
    dataType : i             dataType : i             dataType : i             dataType : i
[p9_speed_o]             [m30_flow]               [v31_settings_o]         [v33a_settings_o]
    controller : PLC28       controller : PLC30       controller : PLC31       controller : PLC33
    dataType : i             dataType : o             dataType : o             dataType : o
[p9_ energy _i]          [v30_pressure]           [m32_flow]               [v33b_settings_i]
    controller : PLC28       controller : PLC30       controller : PLC32       controller : PLC33
    dataType : o             dataType : i             dataType : i             dataType : i
[p9_ energy _i]          [v30_settings_i]         [v32_pressure]           [v33b_settings_o]
    controller : PLC28       controller : PLC30       controller : PLC32       controller : PLC33
    dataType : i             dataType : i             dataType : i             dataType : o
[p12_status_i]           [v30_settings_o]         [v32_settings_i]
    controller : PLC29       controller : PLC30       controller : PLC32
    dataType : i             dataType : o             dataType : i
[p12_status_o]
    controller : PLC29
    dataType : o
```

CHAPTER 4

Efficient *k*-Nearest Neighbour Approach Based on Various-Widths Clustering

This chapter describes an unsupervised SCADA intrusion detection system (IDS) operating with unlabeled SCADA data, which is highly expected to contain abnormal data. The separation between normal and abnormal data is the salient part of the training phase of the proposed system. The *k*-Nearest Neighbour (*k*-NN) approach is found in the literature as being the most interesting and best approach for data mining in general (Wu et al., 2008), and in particular it has demonstrated promising results in anomaly detection (Chandola et al., 2009). This is because an anomalous observation will have a neighborhood in which it will stand out, while a normal observation will have a neighborhood wherein all its neighbours will be exactly like it. However, the visit for all observations in a data set to find the *k*-nearest neighbours for an observation x will result in a high computational cost. This chapter describes the *k*NNVWC (*k*-NN based on Various-Widths Clustering) approach, which efficiently finds the *k*-nearest neighbors for each observation in a data set. A novel various-widths clustering method is described and triangle inequality is adapted to prune unlikely clusters in order to accelerate the search process of *k*-nearest neighbours for a given observation. Experimental results demonstrate that *k*NNVWC performs well in finding *k*-nearest neighbours compared to a number of *k*-NN search algorithms, especially for a data set with high dimensions, various distributions, and large size.

4.1 INTRODUCTION

The Exhaustive *k*-Nearest Neighbour (E*k*-NN) (Cover and Hart, 1967) is a powerful nonparametric algorithm that has been extensively used in many scientific and engineering domains, and is one of the most interesting and best algorithms for machine learning (Papadopoulos, 2006; Wu et al., 2008) and pattern recognition (Shakhnarovich et al., 2008). For instance, it achieves good accuracy in unsupervised outlier detections (Angiulli and Fassetti, 2009; Angiulli et al., 2006; Ghoting et al., 2008; Bay and Schwabacher, 2003). In addition, it demonstrates high

SCADA Security: Machine Learning Concepts for Intrusion Detection and Prevention,
First Edition. Abdulmohsen Almalawi, Zahir Tari, Adil Fahad and Xun Yi.
© 2021 John Wiley & Sons, Inc. Published 2021 by John Wiley & Sons, Inc.

classification accuracy in various applications (Shintemirov et al., 2009; Magnussen et al., 2009; Govindarajan and Chandrasekaran, 2010). However, this algorithm requires the checking of all objects in a data set to find the k-NN for any query object.

To reduce the computational time, a number of algorithms have been proposed, in which the objects are structured in a way that can efficiently accelerate the search for k-NN. For instance, tree-based spatial search algorithms are proposed to structure objects that might have more than three dimensions (Fukunaga and Narendra, 1975; Nene and Nayar, 1997; Liaw et al., 2010; McNames, 2001; Sproull, 1991; Friedman et al., 1975; Kim and Park, 1986; Beygelzimer et al., 2006). These algorithms demonstrated a promising improvement on Ek-NN. However, some of them such as the KD-tree suffer from the problem of the dimensionality drawback (Nene and Nayar, 1997). Hence, a number of tree-based spatial search algorithms such as Ball tree (Fukunaga and Narendra, 1975), Cover tree (Beygelzimer et al., 2006), principal axis tree (PAT) (McNames, 2001) and orthogonal structure tree (OST) (Liaw et al., 2010) are proposed. Despite the enhanced performance offered by these algorithms in accelerating the search of k-NN with high dimensional data, there are still several outstanding issues that need to be resolved such as the contraction time of the tree as well as the traversal cost of the pre-created tree to find the upper and lower search bounds. In addition, the search for "true" k-NN within this area adds extra costs and can substantiality decrease the performance, especially when the number of checked objects is relatively large.

Clustering-based approaches (Eskin et al., 2002a; Prerau and Eskin, 2000; Xueyi, 2011) are proposed as alternative choices for the tree-based spatial search algorithms. Fixed-width clustering (FWC) (Eskin et al., 2002a; Prerau and Eskin, 2000) and k-means (Xueyi, 2011) algorithms are used for grouping the training data set into similar clusters, and like tree-based algorithms, the elimination rules are used to eliminate the clusters that cannot have the k-NN for a query object. Unlike tree-based algorithms that often consist of a large number of nodes, especially when a data set is large in size and dimensions, the clustering-based approaches directly group a training data set into a relatively small number of clusters, and therefore the cost of finding the upper and lower search bounds for a query object is low.

Despite the promising performance of clustering-based approaches, the characteristics of the produced clusters such as size, shape, and compactness can be challenging issues in terms of benefiting from triangle inequality, which is used for elimination, in minimizing the search area as much as possible. Therefore, a training data set that has various distributions can be a challenging issue for both k-means and FWC algorithms. This is because the former assigns sparse objects to the closest cluster, and therefore the cluster can end up with a large radius that overwhelmingly overlaps many clusters. Consequently, the triangle inequality approach may not be adequately applied using the clusters' centroids and radii. Although the latter can set a fixed radius, it is challenging to determine the value that suits all distributions in a data set. Since this algorithm is the extended version, more details are given in Section 4.3.

In this chapter, the k-NN approach based on Various-Widths Clustering (kNNVWC) is described to efficiently search for k-NN in data and has various distributions. kNNVWC is an extended version of FWC where a number of various widths are used to cluster the training data set. In the clustering part, two processes are looped sequentially until the criteria are met: *Partitioning* and *Merging* (see Section 4.3). The former starts clustering with a global width (radius) that is learned from data, and recursively clusters each produced cluster whose members exceed the predefined threshold with their own local width. This process continues until there is no cluster whose size exceeds the predefined threshold, while the latter is called if some criteria are met. The triangle inequality is applied to efficiently accelerate the search for k-NN. The effectiveness of kNNVWC is evaluated in comparison with FWC and well-known tree-based spacial algorithms such as KD-tree (Sproull, 1991) Ball tree (Fukunaga and Narendra, 1975) and Cover tree (Beygelzimer et al., 2006). In this evaluation, 12 data sets that vary in size, dimensionality, and domains, are used.

4.2 RELATED WORK

It is an expensive process to scrutinize each object in a given data set to find the k-NN for a query object, especially for a large data set. Therefore, one or more elimination criteria are used to remove the necessity of checking each object. A branch-and-bound strategy is used in tree-based algorithms to find the upper and lower bounds of the potential search area that definitely has the "true" k-NN for a query object. Fukunaga and Narendra (1975) applied a hierarchical clustering method to divide a data set into a number of subsets, and each subset is further recursively divided into a number of subsets. This process continues until the criteria are met. Then, a binary tree is adapted to structure the results of the decomposition process so that the branch-and-bound search algorithm can be used to reduce the number of distance computations. This tree-based approach is called Ball tree. This algorithm can accelerate the search of k-NN even with high-dimensional data, although its performance is influenced by the clustering method used in the decomposition phase. Hence, Omohundro (1989) proposed five construction algorithms to address this issue. Beygelzimer et al. (2006) proposed a method, called Cover tree, that exploits the intrinsic dimensionality of a data set, where the data set is assumed to exhibit some restricted growth regardless of its actual number of dimensions. Thus, this algorithm can perform well with a data set that complies with this assumption.

In a different approach, Friedman et al. (1977) introduced a balanced k-dimensional tree called KD-tree to structure a data set in a balanced binary tree, where the data set is recursively split into two parts along one axis (dimension), where hyperplane decomposition is used instead of the clustering method. An enhanced version of the KD-tree was introduced by Sproull (1991). However, because of the fast contraction time of the KD-tree and its efficiency in finding the k-NN, it is not appropriate for data that has more than 10 dimensions (Nene and Nayar, 1997). Similarly, Kim and Park (1986) proposed an efficient method called the "ordered partition," although it is appropriate only for low dimensions.

To address the drawback of dimensionality, the Principal Component Analysis (PCA) method (Jolliffe, 2002) has been adapted by a number of approaches (Liaw et al., 2010; McNames, 2001; Chen et al., 1995) to reduce the computational time, especially with high-dimensional data where the first few principal components are used to spin the training data set and query objects. Similar to the algorithms discussed previously, the binary-tree method is adapted in these approaches; however, only projection values are used to construct the search tree. Despite the performance improvement introduced by PCA-based approaches to high-dimensional data, the process of PCA adds extra cost to both construction and search times. For instance, the Principal Axis search Tree (PAT) approach in McNames (2001) requires a relatively long computation time for a large data set or a search tree with great depth, while in the Orthogonal Structure Tree (OST) approach (Liaw et al., 2010), it is reduced. On the other hand, the preprocessing time is less in PAT, while it is relatively large in OST, especially when dimensionality increases.

Clustering-based approaches with free tree structures are proposed (Eskin et al., 2002; Prerau and Eskin, 2000; Xueyi, 2011). That is, the training data set is clustered into subsets where each subset does not have to represent one of the actual classes in the data set. Then, triangle inequality is used to eliminate the clusters that cannot possibly have a *k*-NN. Unlike tree structures, the structure in these approaches is flat, and therefore no extra cost is incurred by traversing the internal nodes. However, the high-dimensional data presents a potential problem for all optimized *k*-NN algorithms in both the construction and search phases. Wang (Xueyi, 2011) proposed a free-tree-based structure approach as an alternative to spatial searching algorithms, especially for high-dimensional data. The author adapted the *k*-means algorithm to directly cluster the data into a number of groups, and then the triangle inequality method was applied in the search for a *k*-NN. However, the author explicitly stated that the maximum benefit of triangle inequality relies on the distribution of a data set.

Similarly, Eskin et al. (2002a) and (Prerau and Eskin 2000) adapted fixed-width clustering to efficiently break down a data set into subsets with less overlap in order to maximize the efficiency of the triangle inequality method. Even though this method requires one pass over the data set, its efficiency is also reliant on the distribution of the data. For instance, data with various densities can cause two problems. The use of a large width can produce large clusters from the dense areas in *n*-dimensional space, while a small width can produce a large number of clusters. Consequently, the search for *k*-NN in a flat structure is expensive with a large number of clusters, while the search among few clusters that are large in size is expensive as well. Moreover, the clustering time gradually increases with the number of clusters. Therefore, the use of fixed-width clustering for optimizing the search for *k*-NN with a data set with large size, high dimensions, and various distributions is an open issue.

4.3   THE *k*NNVWC APPROACH

The FWC algorithm and its inability to cluster a data set that has various distributions are discussed here. We introduce the two main parts of *k*NNVWC: the various-widths clustering part and the exploitation of triangle inequality to efficiently find *k*-NN for a query object at a reasonable cost throughout the partitioned clusters.

4.3.1   FWC Algorithm and Its Limitations

The FWC algorithm is a simple and efficient method for partitioning a data set into a finite number of clusters with a fixed radius w. The procedure *FixedWidthClustering* (Lines 40–61) in Algorithm 2 summarizes the steps of the extended FWC algorithm, where one pass is required to partition a data set. Let $X = \{X_1, X_2, \ldots, X_n\}$ be a set of objects in the data set D and $C = \{C_1, C_2, \ldots, C_n\}$ be the produced clusters from D using FWC. When an object X_i is pulled from D, it is assigned to the closest cluster C_i providing that the distance between X_i and $C_i \leq w$ and $|C| > 0$. Otherwise, a new cluster is created C_{i+1} and X_i is set as its centroid. This will continue until all objects X in D have been scanned.

The nature of FWC constrains its use in outlier detection (Portnoy et al., 2001; Oldmeadow et al., 2004) and breaking down the data set into fixed-width clusters (Eskin et al., 2002) where these clusters, which are created based on fixed-radius, may not be meaningful. Similar to our work, Eskin et al. (2002) reduced the computation time of the *k*-NN algorithm using the FWC as a means of breaking down the data set into smaller and adjacent subsets in order to remove the necessity of checking every object in a data set to efficiently find *k*-NN. However, the key issue of using the FWC in breaking down the data set can be illustrated in the following example.

Consider the example given in Figure 4.1(a) and (b). These are the first two principal components of a sample of the original DARPA data set (Mahoney and Chan, 2003b). This sample contains 52488 objects. 241 objects are labelled as attacks while the rest are labelled as normal. To find the 5 nearest neighbours for an object X_i, all the objects n -1 (52487 objects) must be checked although only 5 objects are required. Figure 4.4(a) shows the importance of using FWC to remove the necessity of checking all objects. For instance, the 5-NN for an object X_i in in the cluster C_{12} can be found in clusters $C_{12}, C_9, C_7,$ and C_4. However, the efficiency of this algorithm is decreased if some of the produced clusters are relatively large. For instance, the clusters C_3 and C_2, which are adjacent, encompass more than 87% of the data set. Hence, finding the 5-NN of any object in these clusters requires checking more than 87% of the objects.

Setting a fixed-width parameter with a small value can prevent large clusters from being produced. However, this will result in the following issues: (i) a large number of clusters can be created that increases the computation time of the FWC, which is $\mathcal{O}(nc)$, where n is the number of objects and c is the number of clusters. This is mainly due to the sparsely distributed objects in n-dimensional space, contributing to the creation of a large number of clusters with very few members in each. This fact is highlighted in Figure 4.4(a) where the data objects in clusters $C_8, C_4,$ and C_9 show a very sparse distribution. (ii) The large number of created clusters also increases the computation time of the search for *k*-NN (see Section 4.3.3). The two main elements

---

Algorithm 2   Various-width clustering

---

 1 **Input:** Data
 2 **Input:** β
 3 **Output:** Clusters
 4 *Clusters* $\leftarrow \phi$; *add*(*Clusters*, [*Data, zeros*, 0]);
 5 *finished* $\leftarrow$ 0;
 6 **while** *finished* == 0 **do**
 7 | Partitioning (*Clusters*, β) ;
 8 | Merging (*Clusters*, β) ;
 9 | **if** |LargestCluster (*Clusters*)| $\leq \beta$ **then**
10 | | *finished* $\longleftarrow$ 1

11 **return** [*Clusters*];
12 **Procedure** Partitioning (*Clusters*, β)
13 | $U \leftarrow$ LargestCluster (*Clusters*) ;
14 | **while** |*U.objects*| $> \beta$ **do**
15 | | $w \longleftarrow$ using eq 4.3;
16 | | **if** *(w == 0)* **then**
17 | | | *U.nonPartitioned*(1);
18 | | | *update*(*Clusters, U*);
19 | | | *continue* ;

20 | | $< tmpClusters > \longleftarrow$ FixedWidthClustering (*U, w*);
21 | | **if** ClusterNum (*tmpClusters*) > 1 **then**
22 | | | *remove*(*Clusters, U*);
23 | | | *add*(*Clusters, tmpClusters*);
24 | | | $U \leftarrow$ LargestCluster (*Clusters*) ;
25 | | **else**
26 | | | $w \longleftarrow w - (w * 0.1)$;
27 | | | go to line 20

28 **Procedure** Merging (*Clusters*, β)
29 | *MergingList* $\longleftarrow \phi$/* list of tuples < childClusterID, parentClusterID>   */
30 | **foreach** *U* in *Clusters* **do**
31 | | $j \longleftarrow$ using eq 4.1 and eq 4.2;
            | | /* ID of cluster contained *U*   */
32 | | **if** $j \neq 0$ **then**
33 | | | put $< U.getID, j >$ in *MergingList*;

34 | **while** *MergingList* $\neq \phi$ **do**
35 | | **foreach** *tuple* in *MergingList* **do**
36 | | | $< i, j > \longleftarrow$ *tuple*;
37 | | | **if** !isParent(*MergingList, i*) **then**
38 | | | | MergeClus (*Clusters,i,j*) ;
39 | | | | remove *tuple* from *MergingList*;

40 **Procedure** FixedWidthClustering (*U, w*)
41 | *tmpU* $\leftarrow \emptyset$;*tmpC* $\leftarrow \emptyset$;*tmpW* $\leftarrow \emptyset$;/* A set of clusters and their centroids and widths
            | */
42 | $q = 0$;/* The number of created clusters   */
43 | *Clusters* $\leftarrow \phi$;
44 | **foreach** *object* in *U.objects* **do**
45 | | **if** *q==0* **then**
46 | | | $q+ = 1$;
47 | | | put *object* in $tmpC_q$;
48 | | | put *object* in $tmpU_q$;
49 | | | $tmpW_q \leftarrow 0$;
50 | | **else**
51 | | | $< j, dis > \longleftarrow$ ClosestCluster (*object,tmpU*) ;
52 | | | **if** $dis \leq w$ **then**
53 | | | | put *object* in $tmpU_j$;
54 | | | | **if** $tmpW_q < dis$ **then**
55 | | | | | $tmpW_q \leftarrow dis$;

56 | | | **else**
57 | | | | $q+ = 1$;
58 | | | | put *object* in $tmpC_q$;put *object* in $tmpU_q$;$tmpW_q \leftarrow 0$;

59 | **for** $i \leftarrow 1$ to q **do**
60 | | *add*(*Clusters*, [$tmpU_i, tmpC_i, tmpW_i$]);

61 | **return** *Clusters*;

(a)   (b)

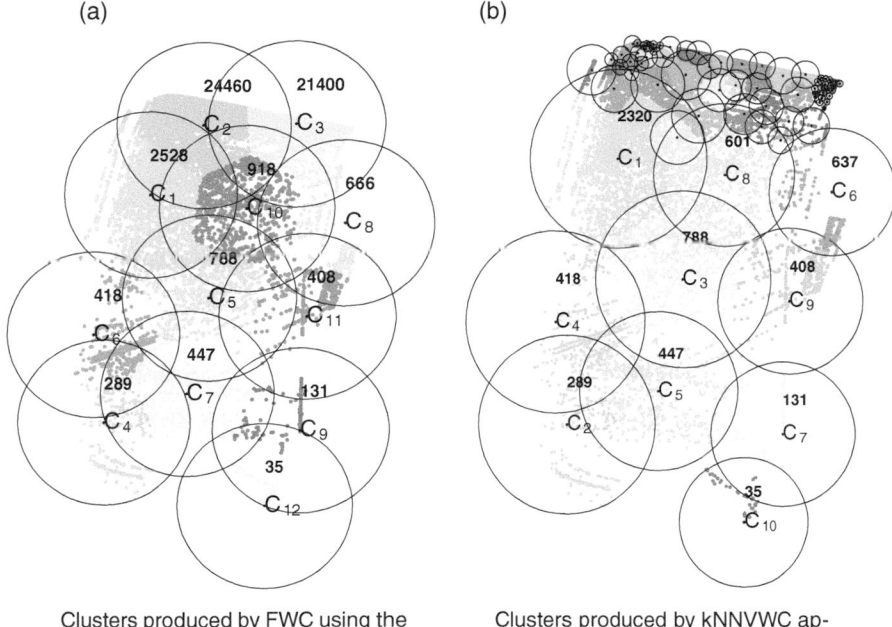

Clusters produced by FWC using the   Clusters produced by kNNVWC ap-
original DARPA data set.   proach the original DARPA data set.

Figure 4.1 Clustering of the two first principal components of a sample of the original DARPA data set.

of *k*NNVWC, namely *various-widths clustering* and *k-NN search*, are elaborated in the following.

4.3.2 Various-Widths Clustering

This section describes the various-widths clustering part, where a data set is partitioned into a number of clusters whose sizes are constrained by the user-defined threshold. Two processes are involved in this operation: *partitioning* and *merging* and they are looped sequentially and executed until the criteria are met. Algorithm 2 shows the steps of the mentioned processes and the variables, data structures, and functions employed by the algorithm are summarized in Tables 4.1 and 4.2.

Partitioning process.   This process partitions a data set into a number of clusters using a large width to resolve the issue of clustering the sparsely distributed objects in *n*-dimensional space. However, large clusters from dense areas will be created such as clusters C_2 and C_3 in Figure 4.4(a). Therefore, each large cluster whose size exceeds a user-defined threshold (maximum cluster size) will be divided into a number of clusters using a width that suits the density of that cluster. This process continues until the sizes of all clusters are less than or equal to the user-defined threshold. Figure 4.1(b)illustrates the produced clusters using various-widths clustering, as can

TABLE 4.1   Variables, data structures, and functions employed by *k*NNVWC

Algorithm 2 Various-widths clustering.

Data	A data set need to be partitioned
Clusters	A list of objects, each object consists of two numeric variables (width and ID) and two arrays that contain members (instances) and centroid of cluster, respectively
β	The maximum cluster size, where any cluster exceeds this threshold will be further partitioned
w	The predefined radius of cluster. The value of this parameter is learned from the data set in (Eq. 4.3)
LargestCluster	This function returns the largest cluster among the set of created clusters
finished	A variable that represents the status of various widths clustering. 0 indicates either further partitioning or merging process is required
Partitioning	A procedure that partitions a data set into a number of clusters using various widths, where the size of each cluster does not exceed the predefined threshold *β*
Merging	A procedure that is responsible for merging the child clusters with their parents
FixedWidthClustering	A procedure partitions a data set into a number of clusters using fixed-width
ClosestCluster	This function computes the distance between the current cluster and others and returns the ID of the closet cluster
isParent	This function is to check that the child cluster, which will be merged into its parent, is not the parent cluster for others
MergeClus	A function that associates the members of a cluster C_i with another C_j and removes cluster C_i from the list of clusters *Clusters*

TABLE 4.2   Variables, data structures, and functions employed by *k*NNVWC

Algorithm 3: Search for *k*-NN throughout partitioned clusters.

Clusters	A list of objects, each object consists of two numeric variables (width and ID) and two arrays that contain members (instances) and centroid of cluster, respectively
k	The number of nearest neighbors need to be looked for
N	A list of IDs of objects and their distances to an object p
Z	An array of IDs of candidate clusters for an object p
NN_k	This function returns *k*-nearest neighbors of an object p in data set d, where $p \notin d$
ClusAscOrder	This function returns a list of IDs of clusters and their distances to the current cluster. This list is sorted by the distances in ascending order

be seen in Figure 4.1(b),where large clusters are partitioned into a number of smaller clusters.

The main steps of this process are summarized in the procedure for *Partitioning* in Algorithm 2. This procedure has two variables: *Clusters* and β. The former is a list of class objects, where each object contains members and properties of a cluster. In the initial step, the whole data set is considered as a cluster and its centroid and width is set with zeros (Line 4).The latter variable is the threshold of the maximum cluster's size. The function *LargestCluster* returns the largest cluster U, that is, not assigned as nonpartitionable, from *Clusters* (Line 13).If the size of U is greater than (or equals) β, Equation 4.3 is used to calculate an appropriate width w for partitioning U. If the value of w equals zero, U is assigned as nonpartitionable (Lines 14–19).This is because the objects in U have similarities in terms of the distance function and therefore they cannot be partitioned. Otherwise, the procedure *FixedWidthClustering* in (Lines 40–61)iscalled to partition U (Line 20).If the number of produced clusters is just one, the value of w is very large and it should be minimized by 10% and used again (Line 26).Otherwise, the new clusters produced from U are added to *Clusters* instead of U, and the largest cluster again is pulled from *Clusters* (Lines 21–24).The steps (Lines 14–26)are repeated until the partitionable largest cluster in *Clusters* is less β.

Merging process.   The partitioning process of a large cluster can lead to the creation of a cluster that is totally contained in another cluster. Therefore, merging such clusters decreases the number of clusters produced, thereby increasing the performance of the search for *k*-NN because the distance computations between a query object and the clusters' centroids will be less when fewer clusters exist. Figure 4.4(a) shows that many objects of the cluster C_2 are located within the range of cluster C_1, but they are associated with the cluster C_2 because it is the closest cluster. To partition the cluster C_2 (as it is large) into a number of clusters, the objects that are located within the range of the cluster C_1 might form a new cluster inside the cluster C_1. Therefore, the merging process is proposed to address this potential problem. Let $C = \{C_1, C_2, \dots, C_n\}$ be the created clusters. The cluster C_i is contained by the cluster C_j, if the following criteria are met:

$$\begin{cases} D(C_i, C_j) + w_i \le w_j & C_i \subset C_j, i \ne j \\ \text{Otherwise} & C_i \subset C_j \end{cases} \qquad (4.1)$$

where $D(C_i, C_j)$ is the distance function between the centroids of clusters i and j, and w_i and w_j are the radii of clusters i and j, respectively. C_i might be contained by a number of clusters $C = \{C_1, C_2, \dots, C_f\}$. Therefore, C_i is merged with the closest C_j that is defined as follows:

$$C_j = \min_{j=1}^{f} D(C_i, C_j) \qquad (4.2)$$

The final step of the merging process involves merging C_i with C_j while C_i is a parent of C_{j+1}. This is undertaken quite simply by firstly merging the child clusters (that are not parents of other clusters) with their parents. The procedure *Merging* in Algorithm 2 summarizes the steps of the merging process. The list of all clusters *Clusters* is iterated to obtain the IDs of child clusters and their respective parents (Lines 28–33).In the next step, all members (objects) of each child cluster are associated with their parent cluster, and the object of this child cluster is removed from the list *Clusters*. To avoid removing any child cluster that might be a parent of other clusters, bottom-up merging is used, where the child clusters that are not parents of any clusters (like leaf nodes in tree structures) are merged first (Lines 34–39).

Parameters.   Two parameters are required in the various-widths clustering, namely the cluster width w and the maximum cluster size β. The selection of these parameters' values have an important impact on kNNVWC's performance. Unlike FWC, cluster width w in various-widths clustering is not sensitive due to the use of the new parameter β and can be derived from the data being clustered. Let D be a data set to be clustered, $NN_k(H_i)$ be the function of k-nearest neighbors for the object H_i, and $clsWidth$ be the function computing the width (radius) of $NN_k(H_i)$, where the width is the distance between the object H_i and the farthest object among its neighbours. The value of k is set to $50\% \times |D|$ to guarantee a large cluster. To find the appropriate global width, we randomly draw a few objects from D, $H = \{H_1, H_2, \dots, H_r\}$ where $r \ll |D|$, and for each randomly selected object, the radius of its k-nearest neighbours is computed and the average is used as a global width for D as follows:

$$w = \frac{1}{r} \sum_{i=1}^{r} clsWidth(NN_k(H_i), H_i)) \tag{4.3}$$

The user-defined threshold β is the second parameter that has an important impact on the performance of kNNVWC because the small value of β will result in a large number of clusters that can decelerate the clustering processing time, in addition to the search for k-NN. Although a large value of β will result in large clusters that can accelerate the clustering process, this will substantially decelerate the search for k-NN. To obtain the near-optimal value of β the size of a data set is considered.

4.3.3   The *k*-NN Search

The influence of this parameter with respect to the size of a data set is investigated here and the suggestion for choosing the near-optimal value is given. This section also shows how triangle inequality can be exploited to efficiently find k-NN throughout the clusters that are produced by Algorithm 2. To eliminate clusters that cannot possibly have k-NN for a query object q, the distance similarity between each cluster and the query object q must be calculated. Prior to discussing the use of triangle inequality for removing impossible clusters, two definitions are presented first.

Definition 4.1 [*k*-nearest neighbors]   The k-Nearest Neighbors (k-NN) of an object p are objects whose distances from p are less than the remaining objects in the

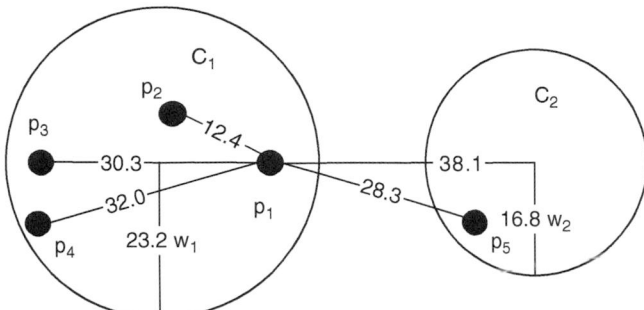

Figure 4.2 An illustration of the use of the triangle inequality for searching nearest neighbors.

data set. Let $X = \{X_1, X_2, \ldots, X_n\}$ be a set of n objects in the data set and $D(X_i, X_j)$ is the distance function between pairs of objects. Given $B = \{B_1, B_2, \ldots, B_k\}$ is a set of k-nearest neighbors of the object p where $B \subset X$, $p \in X$ and $p \notin B$. B is k-NN of p if

$$\max_{n=1}^{k} D(p, B_n) < \min_{n=1}^{|Z|} D(p, Z_n), Z = X \setminus \{B \cup \{p\}\} \text{ is held.}$$

For example, Figure 4.2 shows two clusters C_1 and C_2. The two-nearest neighbors of object p_1 are objects $B = \{p_2, p_5\}$, even though p_5 resides in a different cluster. This is because the object p_5 holds the definition of the two-nearest neighbors of object p_1.

$$\max_{n=1}^{2} D(p_1, B_n) < \min_{n=1}^{|Z|} D(p, Z_n), Z = \{p_3, p_4\}$$

Definition 4.2 [Candidate cluster for an object p]   Let $C = \{C_1, C_2, \ldots, C_n\}$ be the set of clusters and $B_i = \{B_1^i, B_2^i, \ldots, B_k^i\}$ be the k-NN for an object p in a cluster C_i. The cluster C_j is considered as a candidate cluster for the object p, such that $i \neq j$, if the following holds: $D(p, C_j) - w_i < D(p, B_z^i), z = 1, 2 \ldots, k$, where w_i is the radius of the cluster C_j.

This can be illustrated in Figure 4.2. The objects p_2, p_3, and p_4 are the 3-NN of the object p_1 in the cluster C_1. The cluster C_2 is considered as a candidate cluster for the object p_1 because the distance between the object p_1 and the cluster C_2 minus the radius of this cluster is at least less than one of the distances between the object p_1 and its neighbors p_2, p_3, and p_4. For example, $D(p_1, C_2) - w_2 = (38.1 - 16.8) = 21.3 < D(p_1, p_3) = 30.3$ and $D(p_1, p_4) = 32$. Hence, the candidate cluster C_2 might contain objects that are nearer than p_3 and p_4, which is, in this example, the object p_5, and therefore the object p_4 is replaced with it.

Algorithm 3 summarizes the steps involved in a k-NN search. The variables, data structures and functions employed by this algorithm, are summarized in Table 4.1. When a query object q is received, the function *ClusAscOrder* returns a list of cluster IDs and their respective distances to q; the list is named *clus*. The returned data

---

Algorithm 3   Search for k-NN for a query object throughout clusters.

---

1 **Input:** clusters
2 **Input:** k
3 **Input:** q
 /* The query object   */
4 **Output:** N
5 $clus \leftarrow \phi$
 /* list of tuples < clusterID, distance > sorted by
     distance in ascending order   */
6 $clus \leftarrow$ ClusAscOrder $(clusters, q)$
7 $tmpU \leftarrow \phi$
8 **foreach** $\{clusterID, distance\}$ *in clus* **do**
9     $U \leftarrow get(clusters, clusterID)$
10     $tmpU \leftarrow tmpU \bigcup U.objects$
11   remove $< clusterID, distance >$ from $cluID$
12   **if** $|tmpU| > k$ **then**
13   break

14 $T \leftarrow NN_k (p, tmpU, k)$
 /* list of tuples < objectID, distance >   */
15 $N \leftarrow \phi; Z \leftarrow \phi$
16 **foreach** $\{clusterID, cluDis\}$ *in clus* **do**
17     $U \leftarrow get(clusters, clusterID)$
18   **foreach** $\{objID, objDis\}$ *in T* **do**
19   **if** $(cluDis - U.radius) < objDis$ **then**
20   put $clusterID$ in Z
21   break

22 **foreach** $clusterID$ *in Z* **do**
23     $U \leftarrow get(clusters, clusterID)$
24   put $NN_k (p, U.objects, k)$ in N
25 $N = N \cup T$
26 **return** top k-nearest objects from N

---

are sorted by distances in ascending order (Line 6).Atemporal cluster object $tmpU$ is created to represent the closest cluster to q. If the size of $tmpU$ is less than k, the objects of the second closest cluster are merged in $tmpU$. This process continues until the size of $tmpU$ becomes greater than (or equal to) k. Note that all IDs of the joined clusters are removed from the list $clus$ (Lines 7–13).In this step, $|tmpU| > k$, and therefore it is initially assumed to have k-NN for the object q. The function NN_k returns a list T of the IDs of the k-NN from $tmpU$ alongside their cross-posting distances to the object q (Line 14).In fact, the objects in T cannot be guaranteed as "true" k-NN for q. Thus, the candidate clusters, which might contain objects that are nearer than the objects in T w.r.t. the object q, have to be defined from the clusters that have not been merged into $tmpU$. This can be done with fast and simple calculations; no distance computation is required because the distances from the object q to all the clusters and the objects in T have already been computed (see Lines 6 and 14).Therefore, for any cluster whose distance from the object q is d, and the radius w of this cluster is subtracted from $d = d - w$, the cluster is assigned as a candidate cluster when d is less than the distance between the object q and any assumed k-NN objects in T (Lines 15–21).

To find the "true" k-NN for q, the similarity distance between each object in the candidate clusters in Z and the object q is computed. All these objects, in addition to the ones in T, are stored in the list N, and are sorted by their distances to q in ascending order. The final step is to return the top k objects from N as the "true" k-NN for q (Lines 22–26).

4.4 EXPERIMENTAL EVALUATION

This section evaluates the performance of the kNNVWC approach on various data sets. Three aspects are considered in this evaluation: (i) the total number of distance computations, (ii) the computational cost of the construction process, and (iii) the search for k-NN. Four algorithms are compared with kNNVWC. The q-fold cross-validation is applied to each data set, by dividing it into q equal sized subsets. Each time, one of the subsets is used as testing queries, while the remaining $q - 1$ subsets are used as a training data set. Then, the average value of the results of all folds is used as an overall result. In this evaluation, q is set to 10, as suggested by Kohavi (1995) in order to reliably demonstrate the efficiency of any proposed algorithm. All algorithms were implemented in Java and executed under Windows 7 enterprise with 3.4 GHz CPU and 8 GB memory.

4.4.1 Data Sets

Twelve data sets with high, medium, and low-dimensional feature spaces are used for evaluating the proposed approach. These data sets have been selected because they come from various domains and are widely used for the evaluation of data mining methods. Ten of them were obtained from the UCI Machine Learning Repository (Frank and Asuncion, 2013a), while three come from different places. Notably, these data sets are intended to be used for the purposes of classification and outlier detection; however, we use them only to evaluate the performance of the proposed approach to searching k-nearest neighbors in various domains and dimensions. The characteristics of these data sets are briefly described below:

- `arcene`: It contains mass spectra obtained with the SELDI method. This data set has 900 objects, each being described by 1000 features that can be used to separate the cancer patients from healthy patients (Frank and Asuncion, 2013a).

- `SimSCADA`: This data set consists of 12,000 objects, each being described by 113 features. Each feature represents one sensor or actuator reading in the water network systems, which is simulated using the proposed SCADA testbed SCA-DAVT in Chapter 3.

- `multiplefeaturs`. This data set consists of 2000 patterns of handwritten numerals ("0"–"9") extracted from a collection of Dutch utility maps, where each pattern is represented by 649 numeric features (Frank and Asuncion, 2013a).

- `arrhythmia`: This data set consists of 452 objects, each represented by 279 parameters (features) of ECG measurements. It is used to classify a patient into one of the 16 classes of cardiac arrhythmia (Frank and Asuncion, 2013a).

- `gasSensors`: This data set contains 13,910 measurements that come from 16 chemical sensors, and each feature vector contains the eight features extracted from each particular sensor. This results in a 128-dimensional feature vector (Vergara et al., 2012; Rodriguez-Lujan et al., 2014).

- `spambase` This data set contains two classes: spam and non-spam e-mails. It consists of 4601 objects (e-mails), each being described by 57 continuous features denoting word frequencies (Frank and Asuncion, 2013a).

- `kddcup99`: This set comes from the 1998 DARPA Intrusion Detection Evaluation Data (Mahoney and Chan, 2003b).

- `waveform`: It consists of 5000 objects having 40 continuous features, some of which are noise (Frank and Asuncion, 2013a).

- `DUWWTP`: It comes from the daily measures of sensors in an urban waste water treatment plant, and it consists of 527 objects, each represented by 38 features (Frank and Asuncion, 2013a).

- `shuttle`: This data set consists of 43,500 objects, each represented by nine numerical features. It is used in the European StatLog project to compare the performances of machine learning, statistical, and neural network algorithms on data sets from real-world industrial areas (Frank and Asuncion, 2013a).

- `slices`: This data set consists of 53,500 computed tomography (CT) images scanned from 74 different patients (43 male, 31 female). Each CT image is described by two histograms in polar space from which 384 features are extracted (Frank and Asuncion, 2013a).

- `MORD`: This data set consists of 4690 objects, each represented by two features (e.g., temperature and humidity). It is collected from a real wireless sensor network and used for outlier detection purposes (Suthaharan et al., 2010).

4.4.2   Performance Metrics

The key problem of Ek-NN is that it requires a lengthy computation time because each object in the data set must be checked to find the k-NN for a query object. Therefore, the efficient search for k-NN might be the main aim of any optimized algorithm. To measure this efficiency, the construction (if a preprocessing step is required) and search times and the number of distance computations should be considered in the evaluation step.

Reduction Rate of Distance Computations.   This metric is platform-independent where the number of distance computations is not affected by high/low platform resources, although it is data-dependent. To measure the reduction rate of distance computations (RD) for an algorithm Ω, the Ek-NN is used as the baseline. The number of distance computations for Ek-NN is fixed for any k, which is $q \times n$, where n is the number of objects in the data set and q is the number of query objects. Then the performance of Ω for finding k-nearest neighbors in m distance computations is calculated as follows:

$$RD = 1 - \frac{m}{q \times n} \tag{4.4}$$

Reduction Rate of Computation Time.   This metric is platform- and data-set-dependent. The reduction rate of computation time (RC) for Ek-NN is the same for any k because all objects in the training data set will be checked. Then, the RC for an algorithm Ω that takes the time t_1 is calculated as follows:

$$RC = 1 - \frac{t_1}{t_2} \tag{4.5}$$

where t_2 is the time taken by Ek-NN.

4.4.3   Impact of Cluster Size

In the described approach, the number of produced clusters is influenced by the parameter β, which is the maximum size of any produced cluster. Obviously, as shown in Figure 4.3, this parameter can influence the performance of the proposed approach. As a large value of β can result in a small number of large-sized clusters, a large number of objects will be checked to find k-NN. This is illustrated in Section 4.3.1. On the other hand, a small value can result in a large number of small clusters, and this can minimize the search boundaries, especially when k is small. However, the distances between each query object and all produced clusters are required for each object, which adds extra cost in addition to distance computations within the minimized search boundaries.

Clearly, a near-optimal performance relies on the optimal value of β. In this section, we demonstrate how a near-optimal value of β can be inferred from the size of a data set. Since the proposed approach inherits the nature of the FWC algorithm, the exact number of produced clusters cannot be determined; therefore, we tested various values of β to produce a number of clusters that ranges approximately from $0.5\sqrt{s}$ to $8\sqrt{s}$, where s is the size of the data set. In this investigation, data sets varying in terms of domain, size, and dimensionality are used.

As can also be seen in Figure 4.3, the performance of the proposed approach is not very sensitive to the number of produced clusters within the range of $\sqrt{s}$ and $3\sqrt{s}$.

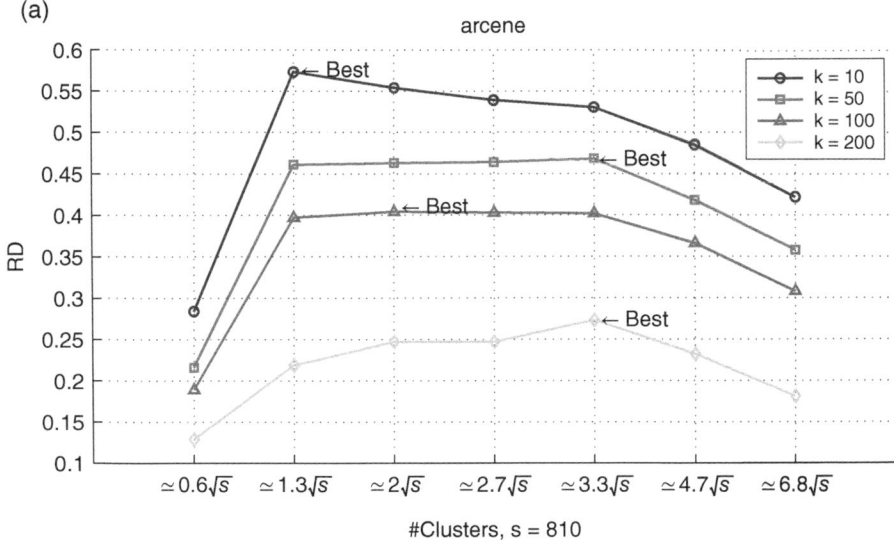

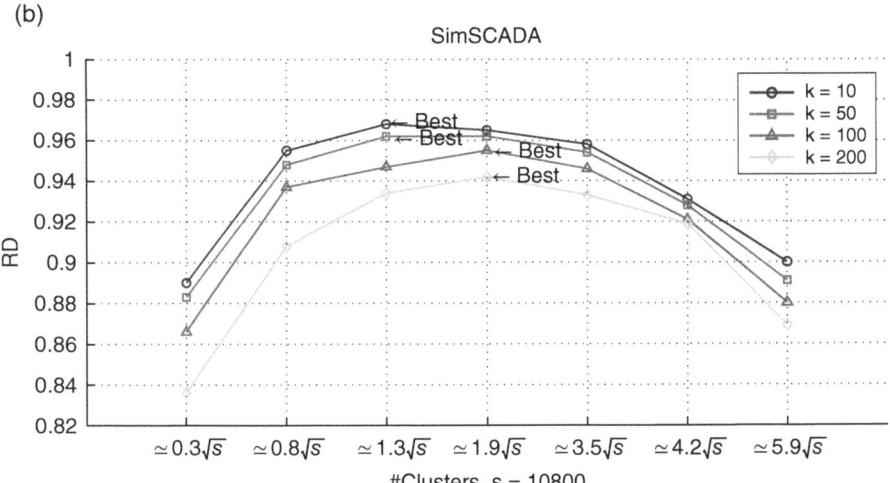

Figure 4.3   An investigation of the impact of cluster size, which is influenced by the value of the parameter β, on the performance of the proposed approach, where s is the size of the training data set.

Moreover, it can be seen that the variance of the performance for each k within this range is small. Therefore, we suggest using a value of the parameter β that produces clusters within this range. However, there might be some data sets that do not comply with this. Thus, a user can find the near-optimal value of β with few attempts.

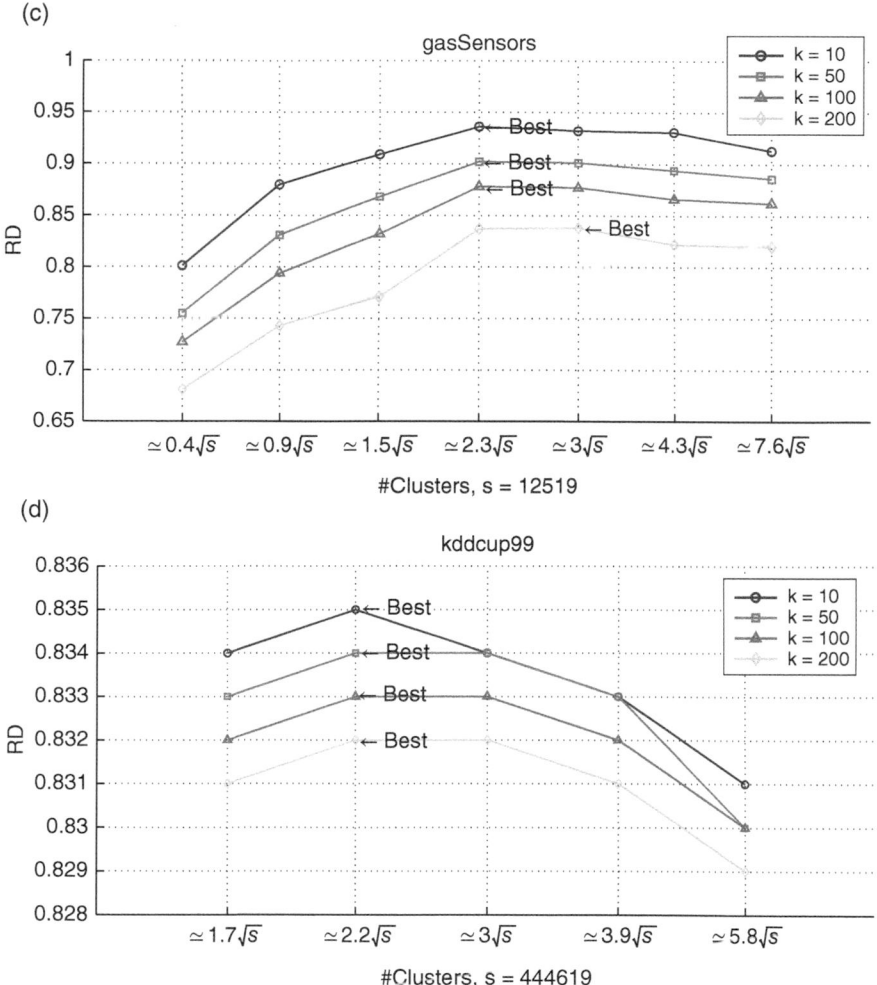

Figure 4.3   (*Continued*)

4.4.4   Baseline Methods

*k*NNVWC is compared here with well-known methods that speed up a *k*-NN search, including: KD-tree, Ball tree, Cover tree, and FWC.

KD-tree.   KD-tree is an algorithm used to structure objects in a binary tree in order to speed up the query time of *k*-nearest neighbor (Sproull, 1991). Because the objects at each node are recursively partitioned into two sets by splitting along one axis (dimension), the contraction time of a KD-tree is very fast. However, the efficiency of the query time decreases when the number of dimensions, *m*, increases, and it is worthwhile when $m \leq 10$ (Nene and Nayar, 1997).

Ball tree.   Ball tree defines k-dimensional hyperspheres (balls) that cover the data objects and structure them into a binary tree that consists of a finite number of nodes, where each node represents a set of objects. The root node represents the whole objects. There are two types of nodes: a leaf node and a non-leaf node. The leaf node contains a list of objects represented by the node; a non-leaf node has two children nodes whose centroids are assigned as follows. The centroid of the objects in the parent node is calculated. Then, the farthest object from this centroid is assigned as the centroid of the first child, while the farthest object from the centroid of the first child is assigned as the centroid of the second child. Afterwards, each object is assigned to the closets child node, and this recursively continues until all criteria are met. This structure of the objects and the use of triangle inequality can accelerate the search for k-nearest neighbors. Further technical and theoretical details about this algorithm can be found in references (Fukunaga and Narendra, 1975; Liu et al., 2006; Omohundro, 1989).

Cover tree.   Cover tree is a state-of-the-art method that efficiently finds the k-NN in high-dimensional data. It is based on the assumption that a data set shows certain restricted growth, regardless of its actual number of dimensions. However, it requires a great deal of preprocessing time, especially when dimensionality increases. In addition, the distribution of the data set plays a major role in its effectiveness (Beygelzimer et al., 2006).

FWC.   FWC is a fixed-width clustering algorithm that is used to partition a data set into small subsets using fixed width. See the previous Section 4.3.1. Eskin et al. (2002) adapted this algorithm with the use of triangle inequality to remove the necessity of checking each object in the data set in order to obtain the k-nearest neighbors.

4.4.5   Distance Metric

To determine the search area that definitely has a k-NN, several computations are required for all baseline methods as well as kNNVWC. In tree-structure-based methods, some computation is required to traverse the tree to find the leaf nodes that have objects within the search area; in flat-structure-based methods (e.g., FWC and kNNVWC), only distance computations between a query object and cluster centroids are required first. Due to the variation of these precalculations, only the total number of distance computations between a query object and the objects in the search area is counted. That is, only the total number of visits for objects is used to locate the k-nearest neighbors for n queries, where n is 10% of the data set. Notably, the Cover tree algorithm is excluded in this comparison because the library, which is the weka jar file (Hall et al., 2009), does not provide a public method for obtaining the number of visited objects in this algorithm. In Section 4.4.6, however, kNNVWC is compared with the Cover tree algorithm based on the computation and construction times that are the ultimate goals of the end-users.

Various values of k, 10, 50, 100, and 200, are tested in this experiment. These values are chosen to compare the performance on small, medium, and large k values.

As previously discussed, the distance computations for Ek-NN are fixed for all values of k. Therefore, all the results are compared with Ek-NN, and demonstrate the amount of deduction that each algorithm can achieve.

Table 4.3 shows the results of the comparison of five algorithms whose total distance calculations are compared with Ek-NN, which represents the worst case where all objects in a data set are visited to find the k-NN for a query object. As previously discussed, the best algorithm according to this comparison is the one that minimizes as much as possible the total number of distance calculations. From these results, we can see that kNNVWC performed the best on the highest dimensional data set *arcene*, which has 10,000 features, while FWC ranks second. However, the large values of k could degrade their performance. This can be attributed to the small size of the data set where all optimized algorithms will visit all objects when the value of k is set equal to the data set size. On the other hand, tree-structure-based methods, such as Ball tree and KD-tree, did not show any improvement with this data set. This is because the search boundaries for each query object overlapped with most nodes in the pre-created tree. Hence, all objects in such nodes must be checked. Similarly, kNNVWC competed against all baseline methods with high dimensional data sets such as *SimSCADA*, *multiplefeaturs*, *gasSensors*, and *Slice*, except for the data set *arrhythmia* with which all methods failed. The reason for this is the small size of this data set. Thus, the Ek-NN method is more appropriate for any small data set.

As for the medium-dimensional data sets *spambase*, *waveform*, *DUWWTP*, and *kddcup99*, the overall performance of kNNVWC is still the best. However, it can be seen that all methods are worse on the data set *waveform* even though its size is relatively large. This is because this data set has a dense data, which might form one single condensed cluster, and therefore the range of the search area for any query object overlaps with the majority of data in n-dimensional space. In this case, a large number of objects must be checked for each query. Clearly, all methods performed well on *kddcup99*. However, due to this, the data set has several large clusters, each of which contains objects that have identical similarities in terms of the Euclidean distance and kNNVWC could not split these large clusters. Thus, the search for k-NN for a query object that is close to one of these clusters will be expensive. This is because all objects in such large clusters will be checked. In fact, kNNVWC would certainly have shown a more significant result if these clusters had been removed.

For low-dimensional data sets *shuttle* and *MORD*, there is an interesting competition among all methods. Even though interesting results are produced by all methods, it is suggested that the KD-tree will be used when dimensions < 10.

4.4.6 Complexity Metrics

In this comparison, kNNVWC is evaluated against the four baseline methods. Two salient aspects of this evaluation are considered: *construction time* taken by each method to build its own structure and the *search time* taken to find k-NN for a number of query objects.

TABLE 4.3 The average reduction rate of distance calculations against E*k*-NN

Data set	K	EkNN	k-d Tree	Ball tree	FWC	*k*NNVWC
arcene	10	7.29×10^4	0.00%	5.97%	53.71%	**61.57%**
	50		0.00%	2.03%	41.54%	**50.28%**
	100		0.00%	0.67%	31.95%	**41.87%**
	200		0.00%	0.06%	12.72%	**25.31%**
SimSCADA	10	1.30×10^7	88.57%	93.45%	91.82%	**97.86%**
	50		74.61%	92.46%	91.52%	**97.56%**
	100		55.39%	90.65%	89.57%	**96.45%**
	200		44.79%	87.22%	85.96%	**94.63%**
multiplefeaturs	10	3.60×10^5	0.00%	0.02%	55.27%	**70.51%**
	50		0.00%	0.00%	47.54%	**63.91%**
	100		0.00%	0.00%	41.73%	**59.24%**
	200		0.00%	0.00%	38.93%	**55.56%**
arrhythmia	10	1.83×10^4	0.00%	0.00%	4.43%	**6.56%**
	50		0.00%	0.00%	3.24%	**5.94%**
	100		0.00%	0.00%	2.48%	**5.86%**
	200		0.00%	0.00%	1.59%	**5.16%**
gasSensors	10	1.74×10^7	90.47%	78.61%	91.59%	**95.25%**
	50		76.91%	71.97%	88.18%	**91.78%**
	100		68.46%	67.82%	85.69%	**89.30%**
	200		59.71%	62.85%	82.56%	**85.11%**
spambase	10	1.90×10^6	57.80%	3.43%	82.39%	**90.87%**
	50		40.73%	1.61%	80.24%	**86.79%**
	100		35.16%	0.13%	78.63%	**83.68%**
	200		28.81%	0.10%	76.03%	**78.53%**
waveform	10	2.25×10^6	0.00%	0.00%	2.47%	**3.08%**
	50		0.00%	0.00%	0.89%	**1.39%**
	100		0.00%	0.00%	0.08%	**0.74%**
	200		0.00%	0.00%	0.01%	**0.43%**
DUWWTP	10	2.51×10^5	0.14%	0.00%	**5.86%**	4.50%
	50		0.00%	0.00%	**2.98%**	2.29%
	100		0.00%	0.00%	2.02%	**2.06%**
	200		0.00%	0.00%	1.20%	**1.86%**
shuttle	10	1.70×10^8	**99.51%**	98.20%	95.39%	95.71%
	50		**98.57%**	97.55%	92.89%	92.30%
	100		**97.79%**	97.07%	90.98%	89.79%
	200		**96.79%**	96.37%	88.52%	86.33%
MORD	10	1.98×10^6	**97.84%**	94.15%	91.20%	94.76%
	50		**96.44%**	93.18%	90.08%	93.56%
	100		**94.88%**	90.50%	88.82%	92.06%
	200		**92.11%**	88.02%	85.94%	89.78%
Slice	10	2.58×10^8	24.38%	47.81%	73.66%	**89.23%**
	50		4.48%	21.07%	65.26%	**83.22%**
	100		1.90%	15.88%	61.72%	**80.73%**
	200		0.76%	12.31%	57.84%	**78.23%**
kddcup99	10	2.20×10^{10}	78.38%	78.00%	73.91%	**83.87%**
	50		78.32%	77.90%	73.83%	**83.82%**
	100		78.27%	77.84%	73.79%	**83.76%**
	200		78.19%	0.00%	73.73%	**83.67%**

Search Time. This part provides an evaluation of the efficiency of each method against EkNN in searching for k-NN. Note that the efficiency value τ is calculated as in Section 4.4.2 and this value is categorized into three statuses (Better, Similar, and Worse) as follows:

$$\begin{cases} 0 < \tau < 1 & \textit{Better} \\ \tau = 0 & \textit{Similar} \\ \tau < 0 & \textit{Worse} \end{cases} \qquad (4.6)$$

That is, the Ω method is better when it takes less time than EkNN to find k-NN and a similar status when its search time is exactly the same as the time consumed by EkNN. Otherwise, the status is worse, and this happens when the search time is longer than the time taken by EkNN. Therefore, any method producing a result greater than zero can be an alternative for EkNN. However, the method with the highest efficiency value is the best choice. On the other hand, EkNN is the best when $\tau \leq 0$ because it does not require any preprocessing step and is simple to implement.

In this evaluation, the same values of k that were used in Section 4.4.5 are also used here. Figure 4.4 shows the efficiency of each method when searching for the k-NN in various data sets. Overall, the tree-based methods KD-tree, Ball tree and Cover tree demonstrate poor results over the high-dimensional data sets, especially when the k value is large, while Cover tree can be the best among them when dimensionality increases. On the other hand, flat-structure-based methods, FWC and kNNVWC, demonstrate promising results except with the small size data set *arrhythmia* and their efficiency is stable even with a large value of k. Nevertheless, for high-dimensional data, kNNVWC is the most efficient method compared with all other methods.

With the medium-dimensional data sets, tree-based methods show good results only for *kddcup99*, with a slightly lower efficiency when dimensionality increases. The relatively large size of this data set is the possible reason for this improvement. This is because EkNN, with which all results were compared, is very expensive using this data set. On using other hand, flat-structure methods demonstrate significant and stable results with all k values except with *waveform* and *DUWWTP*. The explanation for this was previously given in Section 4.4.5; the former is very dense and relatively small, while the latter is very small. However, it is quite obvious that the efficiency value of kNNVWC is greater than the FWC one by approximately 10%. As shown, the overall efficiency of all methods with the low-dimensional data sets is good except for Cover tree, which appears to have low efficiency with large k values.

Construction Time. All methods involve a preprocessing step prior to the k-NN search and all use a clustering-based method for the decomposition process except for the KD-tree, which uses hyperplane separation. This part provides an experimental evaluation of the efficiency of each method in building its own structure.

Figure 4.5 demonstrates the construction time for each of the methods on various data sets. It can be seen that the efficiency of Cover tree and Ball tree in comparison with other methods is influenced by the number of dimensions and the distribution of a data set. For instance, Cover tree exhibits the worst construction time on the

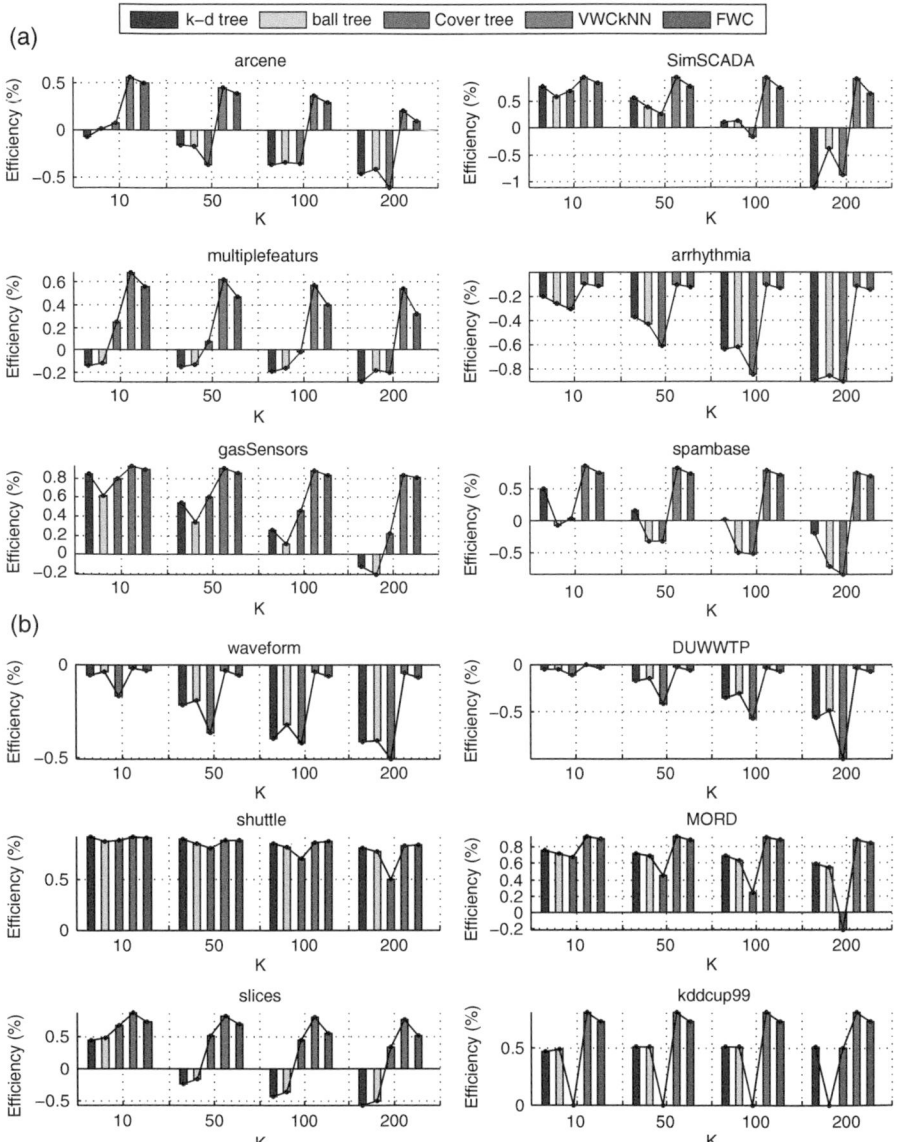

Figure 4.4   The efficiency of the baseline methods and *k*NNVWC against E*k*NN in searching for *k*-NN.

high-dimensional data *arcene*, while Ball tree is relatively acceptable compared to others. The opposite is true for the high-dimensional data *SimSCADA*. This fluctuation can be seen in other data sets. Therefore, this might be a result not only of dimensionality but of distribution. Clearly, FWC is the worst method and is influenced

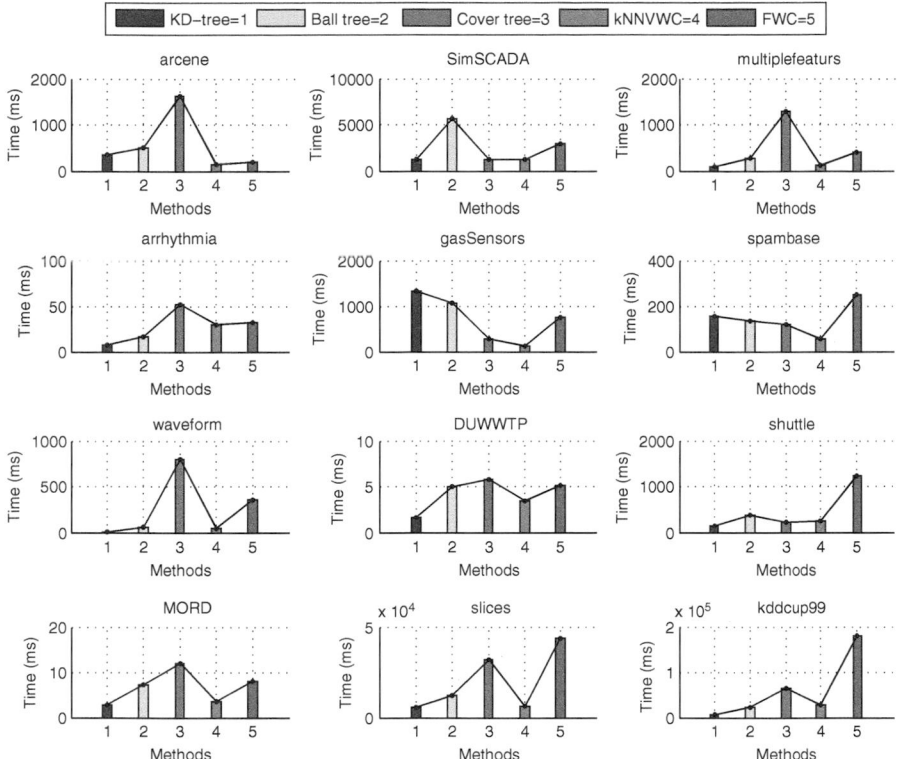

Figure 4.5   The construction time of the baseline methods and kNNVWC for each data set.

by the size of the data set. Overall, kNNVWC and KD-tree appear to be the most efficient and stable methods.

In general, comparing all methods, kNNVWC demonstrates an interesting efficiency in terms of construction and search times, especially with large and high-dimensional data. In addition, it adds only a small extra cost in the worst case scenario when compared with EkNN.

4.5 CONCLUSION

This chapter described in detail kNNVWC, a novel nearest-neighbour search approach based on various-widths clustering. kNNVWC is able to partition a data set using various widths, with each width suiting a particular distribution. Two features have been introduced, which are compared with the use of a fixed width: (i) the balance between the number of produced clusters and their respective sizes has maximized the efficiency of using triangle inequality to prune unlikely clusters; (ii) the clustering with a global width first produces relatively large clusters, and each cluster is independently partitioned without considering other clusters, thereby reducing the clustering time.

In the described experiments, 12 data sets, which vary in domains size, and dimensionality, have been used. The experimental results showed that kNNVWC outperformed four algorithms in both construction and search times. Moveover, the new approach has shown promising results with high-dimensional data in searching for the k-nearest neighbors for a query object.

CHAPTER 5

SCADA Data-Driven Anomaly Detection

This chapter introduces a novel unsupervised SCADA Data-driven Anomaly Detection approach (SDAD), generating from unlabeled SCADA data, proximity anomaly-detection rules based on the clustering method. SDAD monitors the inconsistent data behavior of SCADA data points, and two methods are suggested here: (i) a separation between the consistent and inconsistent observations, which are generated by the data points of a given SCADA system, where kNNVWC is used as the essential part of this process, and (ii) the clustering-based proximity rules for each behavior, whether consistent or inconsistent, are automatically extracted from the observations pertaining to that behavior. The SCADAVT tesbed described in Chapter 3 is used to generate simulated testing data sets that contain the data of tens of SCADA data points. Moreover, real SCADA data sets are also used to evaluate SDAD. Experiment results show that SDAD has significantly better accuracy and efficiency in comparison with existing clustering-based intrusion detection approaches.

5.1 INTRODUCTION

The evolution of SCADA data can reflect the system's state: either consistent or inconsistent, and therefore the monitoring of the evolution of SCADA data for a given system has been proposed as an efficient tailored IDS for SCADA (Carcano et al., 2011; Fovino et al., 2012; Rrushi et al., 2009b; Zaher et al., 2009; Alcaraz and Lopez, 2014; Alcaraz and Lopez). Two attack scenarios from Jared Verba (2008) are used to motivate the use of SCADA data to build anomaly detection models for monitoring the behavior of SCADA systems. Figure 5.1 illustrates that when an attacker compromises the front end processor (FEP), three actions can be performed as follows: (i) initializing connection with remote terminal unit ($RTU_{1.1}$) and sending a command without receiving a corresponding command from the application server, (ii) dropping the command sent from the application server to $RTU_{1.1}$ and frogging feedback information, sent back to the application server, to meet his/her attack, and

SCADA Security: Machine Learning Concepts for Intrusion Detection and Prevention,
First Edition. Abdulmohsen Almalawi, Zahir Tari, Adil Fahad and Xun Yi.
© 2021 John Wiley & Sons, Inc. Published 2021 by John Wiley & Sons, Inc.

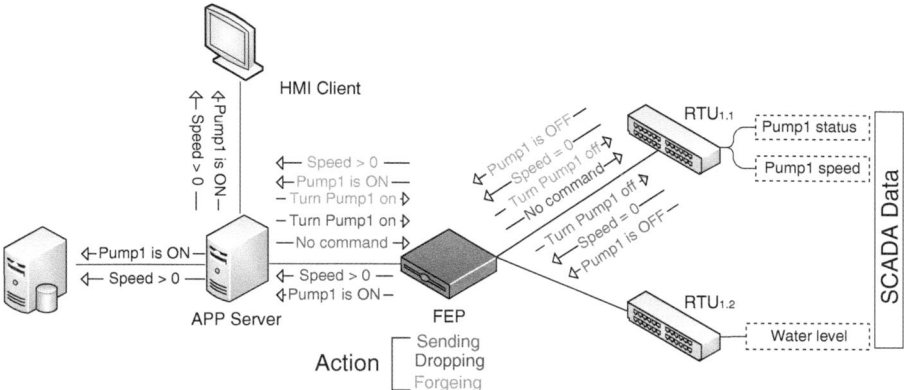

Figure 5.1 Compromised FEP sends undesired command and falsifies the feedback information.

(iii) frogging the command sent from the application server to $RTU_{1.1}$ and also frogging feedback information sent back from $RTU_{1.1}$ to the application server. In fact, all commands sent to $RTU_{1.1}$ will be trusted because they are syntactically valid and sent from the FEP.

In this scenario, two inconsistent data emerged: an inconsistent network traffic pattern and inconsistent SCADA data. The inconsistent network traffic pattern can be identified as follows. (i) FEP is not an intelligent device that can make a decision and send a command to $RTU_{1.1}$ without receiving a corresponding command. (ii) The dropped command at FEP will be shown up in the network stream from the application server to the FEP but not in the network stream from the FEP to the $RTU_{1.1}$, while the frogged commands between the application server and $RTU_{1.1}$ can be identified by the inconsistent SCADA data. For example, the command in the network stream from the application server to the FEP shows that the status of pump1 is ON, while in the network stream from the FEP to the $RTU_{1.1}$ is OFF. Clearly, watching for inconsistencies in commands in this scenario helps to detect that the aforementioned MITM attacks are performed from FEP. However, in the following, we demonstrate a scenario where monitoring of inconsistencies in commands will fail to detect MITM attacks.

Figure 5.2 illustrates that when an attacker compromises the application server, which is intelligent and can initiate independent actions, and also drop commands sent from the operator, an unsafe condition could be created. In the same Figure 5.2, the attacker initializes the command from the application server to the turn of pump1, and it can be seen that the network traffic stream and SCADA data between $RUT_{2.1}$ and the application server are consistent for this command. However, the SCADA data such as the speed and status of *pump*1 could be inconsistent with the sensory node of water level in $RTU_{2.2}$, as they are set with values that violate the specifications of the system from an operational perspective.

As mentioned previously, the monitoring of the behavior of SCADA systems through the evolution of SCADA data has attracted the attention of researchers. Jin

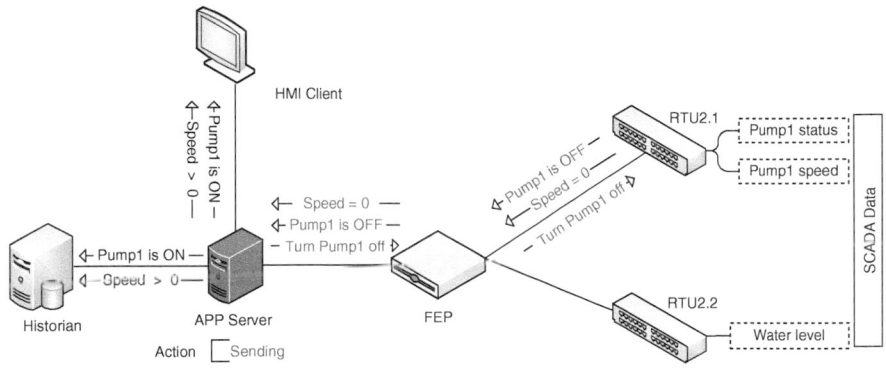

Figure 5.2 Compromised application server sending false information.

et al. (2006) extended the set of invariant models with a value range model to detect any inconsistent value for a particular data point. Similarly, Alcaraz and Lopez (2014) monitor each data point (sensory node) individually using a predefined threshold (e.g., min, max), and any reading that is not inside a prescribed threshold is considered as an anomaly. These approaches are good for monitoring one single data point. However, the value of an individual data point may not be abnormal, but in combination with other data points it may produce an abnormal observation, which very rarely occurs. To address this issue, analytical approaches in Fovino et al. (2010a), Carcano et al. (2011), and Fovino et al. (2012) were proposed to identify the range of critical states for multivariate data points whereby any inconsistent state of the analyzed data points can be monitored. Similarly, Rrushi et al. (2009b) applied probabilistic models to estimate the normalcy of the evolution of multivariate data points. However, pure "normal" training data is required for the probabilistic approach, and this requires the system to operate for a long time under normal conditions. However, this cannot be guaranteed, and therefore any anomalous activity occurring during this period will be learned as normal. Moreover, analytical approaches require expert involvement to analyse and identify the inconsistent range of data points. This, however, results in time-expensive processing and is prone to human errors.

This chapter describes a novel SCADA unsupervised Data-driven Anomaly Detection approach (SDAD), generating from unlabeled SCADA data, proximity anomaly detection rules based on a clustering method. SDAD is intended to monitor the inconsistent data behavior of SCADA data points. SDAD efficiently separates inconsistent from consistent observations in the learning data set for multivariate data points; in addition, the proximity detection rules for each behavior, whether consistent or inconsistent, are automatically extracted. Moreover, SDAD works in an unsupervised mode where purely "normal" training data is not required, and it does not require expert involvement to extract detection rules. This will help to reduce time-expensive processing as well as eliminate human errors. Seven different data sets are used to evaluate the effectiveness of SDAD; two are generated using the SCADA testbed

SCADAVT, described Chapter 3, while the other five are real data sets that consist of both consistent/inconsistent observations of data points.

This chapter is organized as follows. Section 5.2 provides a characterisation of consistent/inconsistent observations for point data and the details of the proposed approach. Section 5.3.5 presents the experimental set-up, followed by results and analysis in Section 5.4. The conclusion about this work is provided in Section 5.6.

5.2 SDAD APPROACH

This section describes consistent/inconsistent observations for SCADA points and the two methods that contribute to the development of SDAD: (i) a method that separates inconsistent observations from consistent ones of multivariate SCADA points and (ii) a method that extracts proximity-based detection rules, which is used to detect inconsistent observations. Figure 5.3 shows the different steps of SDAD. The SDAD approach is based on the k-nearest neighbor approach to assign a inconsistency score for each observation produced by the multivariate SCADA points over a period of time. The fast and efficient k-nearest neighbor approach kNNVWC is used as an efficient algorithm to find the k-nearest neighbour observations for each current observation so that we can assign an inconsistency score for that observation based on its neighborhood density.

5.2.1 Observation State of SCADA Points

The data of SCADA points, such as sensor measurement and actuator control data, are data sources for the unsupervised SCADA data-driven IDS. The data consistency of SCADA points represent the normal current system state, while inconsistency indicates malicious actions (Carcano et al., 2011; Fovino et al., 2012; Rrushi et al., 2009b). The consistency is defined by the specifications, which describe the valid (or acceptable) data in terms of the system's operational perspective. From the SCADA testbed presented in Chapter 3, Figure 5.4 shows the normal operation producing consistent observations of the following SCADA points P_{st_1}, P_{st_2}, P_{st_3}, P_{sp_1}, P_{sp_2}, P_{sp_3}, and T_1, where P_{st_i} and P_{sp_i} represent the status and the speed of the $Pump_i$, respectively, and T_j represents the water level in the $Tank_j$.

Definition 5.1 [Observation of SCADA points] It is a combination of data produced by SCADA points such as sensors and actuators at a certain period of time t. An observation for n points can be represented by a vector $p \in R^n$.

Definition 5.2 [Inconsistent/consistent observation] An observation is considered as inconsistent when the anomaly score for the observation based on its neighborhood density significantly deviates from the anomaly scores of all the learning observations. Otherwise, it is considered as consistent.

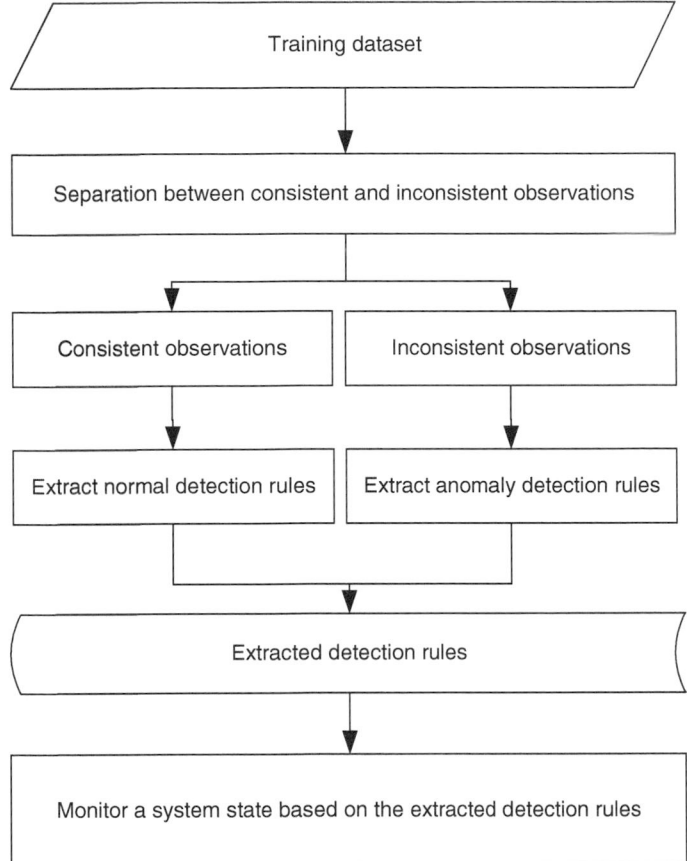

Figure 5.3   The steps of the SDAD approach.

5.2.2 Separation of Inconsistent Observations

As depicted in Figure 5.2, the separation step is the first phase. To perform this with unlabeled data, two assumptions are held: the number of consistent observations vastly outperforms the inconsistent observations and the inconsistent observations data must be statistically different from consistent observations. Therefore, the SDAD approach will be inappropriate for any situation that does not meet these two assumptions. The preliminary investigations show that inconsistent observations have a similar definition of outliers in n-dimensional space and are sparsely distributed in an informal way. That is, they could take various densities of n-dimensional space. Two steps are involved in separating consistent/inconsistent observations as follows.

Inconsistency scoring.   The proposed inconsistency scoring method uses a hybrid of local and global outlier detection approaches (Breunig et al. 1999, 2000; Šaltenis Vydunas 2004; Arning et al., 1996). In the local outlier detection

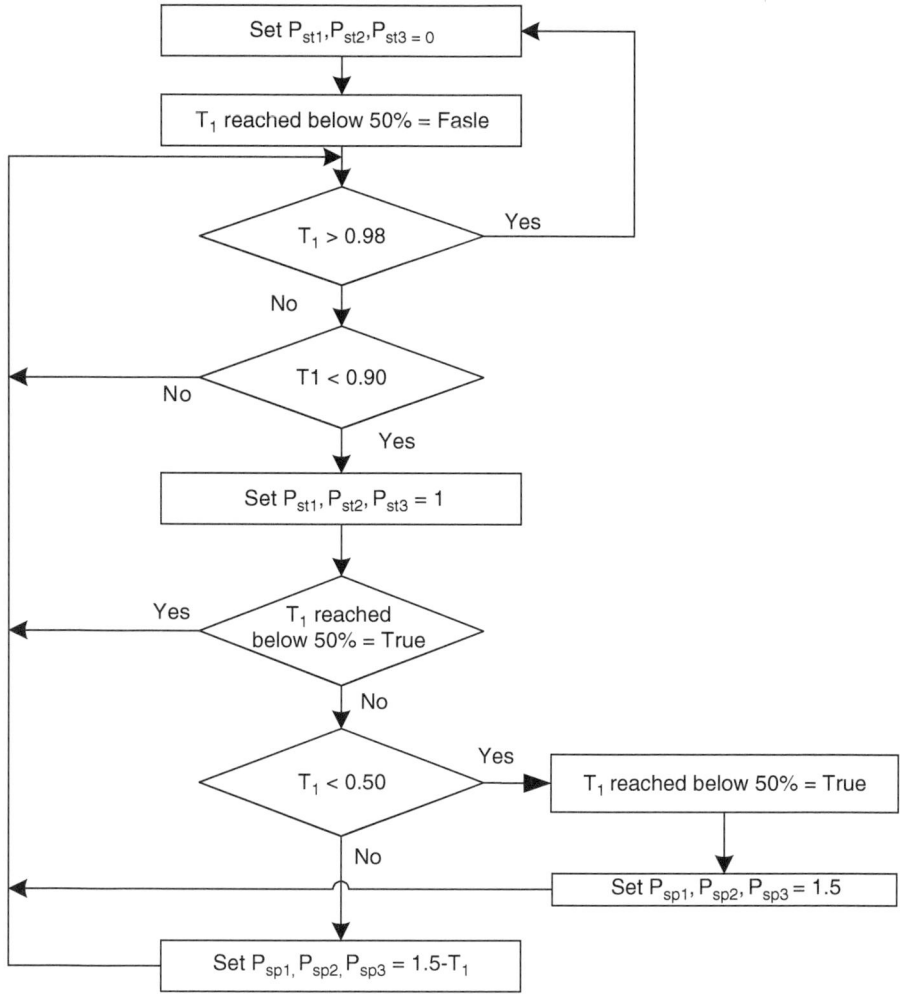

Figure 5.4 The normal operation of the SCADA points $P_{st_1}, P_{st_2}, P_{st_3}, P_{sp_1}, P_{sp_2}, P_{sp_3}$.

approach (Breunig et al. 1999, 2000), only local outliers are detected. This works well in a particular application domain whose normal behavior forms a number of clusters that have different densities. Comparatively, the global approach (Šaltenis Vydunas, 2004; Arning et al. 1996) does not work well when the outliers are contained in the reference points. Since the initial investigation revealed different densities in inconsistent observations in the data space, neither local nor global approaches are appropriate solutions for our problem. The choice of the best approach should not be predominantly influenced by either local or global approaches. Therefore, the inconsistency scoring method will need to rely on an average of distances of the nearest neighbors and the number of neighbors k that play a major role in the influence of local and global approaches. The larger the value of k, the higher the influence of the global approach. For example, when k equals the size of the data set DS, the inconsistency score for an observation s_i is the average distance from s_i to all observations in DS. This is similar to the global approaches proposed in Šaltenis Vydunas (2004) and Arning et al. (1996).

Let D_{points} be a matrix of data points, where $D_{points} = \{d_1, \ldots, d_n\}$. Let X be a vector of values of d_i, $X = \{x_1, x_1, \ldots, x_m\}$. Let S_i denote an observation in D_{points}, $S_i = \{s_{i,1}, s_{i,2}, \ldots, s_{i,j}\}$, $i = 1, \ldots, m$, $j = 1, \ldots, n$, where m is the number of observations and n is the number of data points. Any observation S_i could be either consistent or inconsistent. Therefore, in the identification phase, a set of consistent observations, $R = \{r_1, r_2, \ldots, r_f\}$, and a set of inconsistent observations, $O = \{o_{f+1}, o_{f+2}, \ldots, o_m\}$, are identified, where $R \subseteq D_{points}$, $O \subseteq D_{points}$, $R \cup O = D_{points}$, and $R \cap O = \phi$.

To compute the inconsistency score for each observation S_i, let Ω be a function of Euclidean distance between two observations $s = (s_1, \ldots, s_d)$ and $p = (p_1, \ldots, p_d)$, defined as follows:

$$\Omega(s,p) = \sqrt{\sum_{i=1}^{n}(s_i - p_i)^2} \tag{5.1}$$

where n is the number of data points (attributes). Let k be a positive parameter such that $2 \leq k \leq |DS|$ and Ψ and Υ be the functions of an inconsistency score and the k-nearest neighbors for the observation s_i, respectively. The inconsistency score for an observation is defined as follows:

$$B = \Upsilon(s_i, DS \setminus s_i, k) \tag{5.2}$$

where B is the k-nearest neighbors of s_i:

$$\Psi(s_i, DS, k) = \frac{1}{k}\sum_{j=1}^{k}\Omega(s_i, B_j) \tag{5.3}$$

For example, consider Figure 5.5, where a data set D of 11 observations of $\mathbb{R}^2$ is shown. Let $k = 4$. In this example, the inconsistency score for the observation O_1 based on the Euclidean distance is computed as follows.

We find the four-nearest neighbors of O_1 using the function Υ:

$$\Upsilon(O_1, D \setminus O_1, 4) = \{s_3, s_5, s_6, s_7\}$$

Then, the inconsistency score for the observation O_1 is computed as the average distance from O_1 to $\{s_3, s_5, s_6, s_7\}$:

$$\Psi(O_1, D, 4) = \frac{27.7 + 22.4 + 24.6 + 20.3}{4} = \frac{95}{4} = 79.775$$

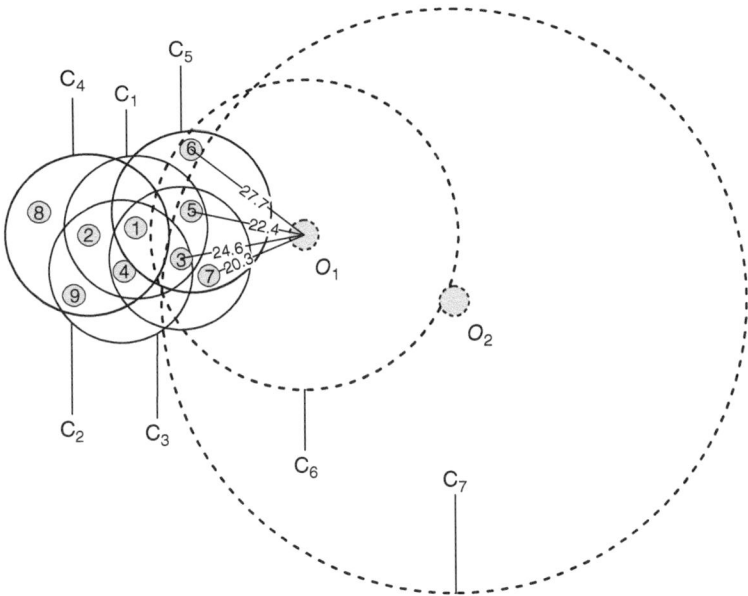

Figure 5.5   Illustration of an inconsistency scoring method based on intra-cluster cohesion factor.

where, the numerators represent the Euclidean distances of the four-nearest neighbors of O_1. Note that the approach kNNVWC that has been proposed in Chapter 3 is used to efficiently to find the k-nearest neighbors.

Figure 5.5 illustrates the k-nearest neighbors of the observation s_i. For example, the k-nearest neighbors of the observation 1, which is a consistent observation, with respect to $k = 5$ are 2, ... , 5. This represents a cluster c_1 whose centre and size are observations 1 and k, respectively.

The inconsistency score of the observation 1 is measured by the average distance of all observations 2, ... , 5 in c_1 to observation 1. It can be seen that the clusters c_1, c_2, and c_3 in Figure 5.5 have similar radii, which suggests that their centroid observations may have a similar inconsistency score. In fact, this is not always true because the inconsistency score is computed by the inter-compactness of the cluster and not by the reachability distance Ankerst et al. (1999). Since the observations 2 and 5 are centroids of clusters whose radii are larger than c_1, c_2, and c_3, they will obtain higher inconsistency scores. The radii of the clusters c_6 and c_7 whose centroids are given by the inconsistent observations O_1 and O_2, respectively, are clearly the largest radii, and are considered as the inconsistent observations because they have relatively large radii that can significantly deviate from the mean of the radii of other observations. Algorithm 4 summarizes the steps used to calculate the inconsistency scores for each observation s_i.

The Separation Threshold.   Since all observations are assigned with inconsistency scores, an appropriate threshold η to determine whether the observation is

---

Algorithm 4   An inconsistency scoring algorithm

---
1 **Input:** DS
  /* A matrix of unlabelled SCADA data consisting of
     m observations and n attributes   */
2 **Input:** k
  /* # of nearest neighbors   */
3 **Output:** IncList
  /* list of inconsistency scores   */
4 $IncList \longleftarrow \emptyset$;
5 **for** $i \leftarrow 1$ **to** $|DS|$ **do**
6   $\lfloor IncList[i] = \Psi(DS[i], DS, k)$;
7 **return** [IncList];

---

consistent or inconsistent is required. In fact, the selection of the near-optimal threshold plays an important role in supporting the robustness of the inconsistency scoring method, while an inappropriate threshold leads to bad results regardless of the criticality scoring method. It is obvious that labeling consistent observations as inconsistent observations will result in a high false positive rate. Moreover, tuning the threshold to reduce the false positive rate is a critical operation because a number of true inconsistent observations will be missed. In the anomaly detection methods that are based on anomaly scoring, observations are sorted in descending order. An analyst might choose to either analyze the top few outliers or use a cut-off threshold to select the anomalies. Therefore, based on the assumption that anomalies constitute a tiny portion of the data, the top small percentage (say between 1% and 5%) of the observations that have the highest anomaly scores, are assumed as anomalies. Figure 5.6 shows the critical step in selecting the best threshold.

It can be seen that the best threshold for the data set *SIRD* is the top 3% of the observations that have the highest anomaly scores, while the best threshold for the data set *MORD* is 0.01. However, under the aforementioned assumption, and in addition to the empirical investigation of the used data sets, it is highly possible to attempt a range of thresholds from 0.50% to 5% increased by 0.50%. That is, the threshold parameter η can be increased (or decreased) by 0.50% until good results are obtained.

5.2.3   Extracting Proximity-Detection Rules

The various constraints of SCADA systems (e.g. real-time nature, lack of memory resources, limited computation power) require a tailored IDS that monitors a target system in at least near real time and operates in a constrained-resource environment. In practice, feeding the IDS with the identified consistent and inconsistent observations for monitoring is not a practical approach because: (i) a large memory capacity is needed to store all training observations and (ii) it is time-consuming

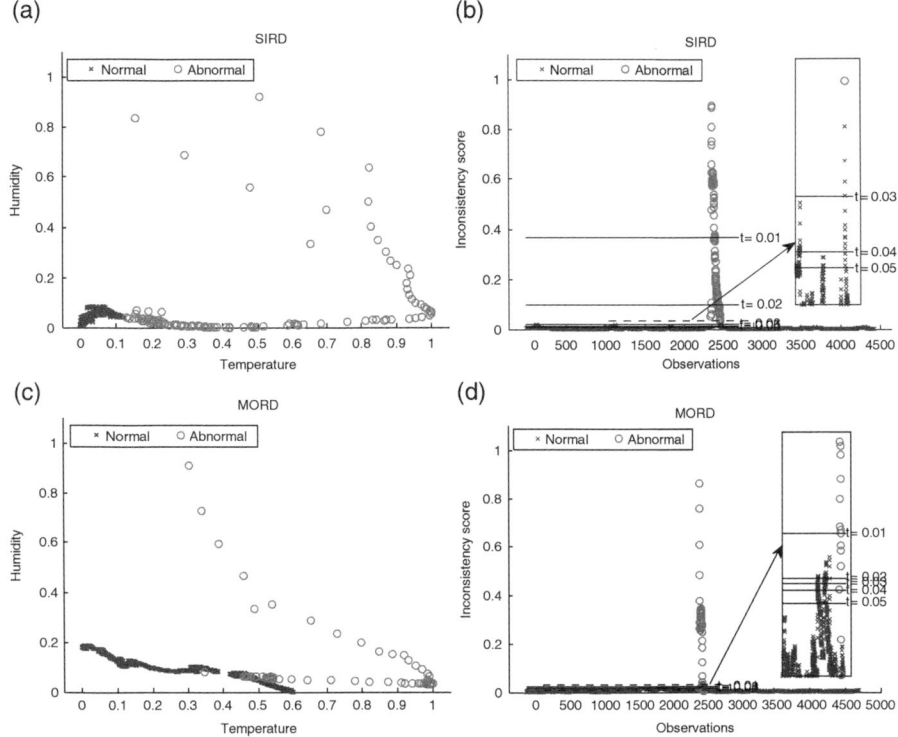

Figure 5.6   (a and c) The behavior of consistent/inconsistent observations of two process parameters in two-dimensional space and (b and d) inconsistency scores after applying inconsistency scoring algorithm.

to calculate the similarity between the current observation and each learned observation in the monitoring phase. This section describes the detection rule extraction method used to extract a few detection rules that fully represent the entire identified observations.

As shown in Figure 5.3, the extraction of proximity-detection rules comes after the separation phase. The set of consistent observations is denoted as $R = \{r_1, r_2, \ldots, r_k\}$ and the set of inconsistent observations is denoted as $O = \{o_{k+1}, o_{k+2}, \ldots, o_m\}$, where $R \cap O = \phi$.

It is assumed that the consistent observations of data points will create dense areas and will constitute a large portion of a training data set, while the inconsistent observations will be sparsely distributed in the n-dimensional space and constitute a tiny portion. Hence, R has one or more high-density clusters. As we are mainly interested in a few extracting detection rules, which can represent the built detection models, the fixed-width clustering method (Portnoy et al., 2001) is adapted to cluster the consistent and inconsistent observations into microclusters with a constant fixed width. However, choosing the appropriate fixed-width value is a challenging task. This is

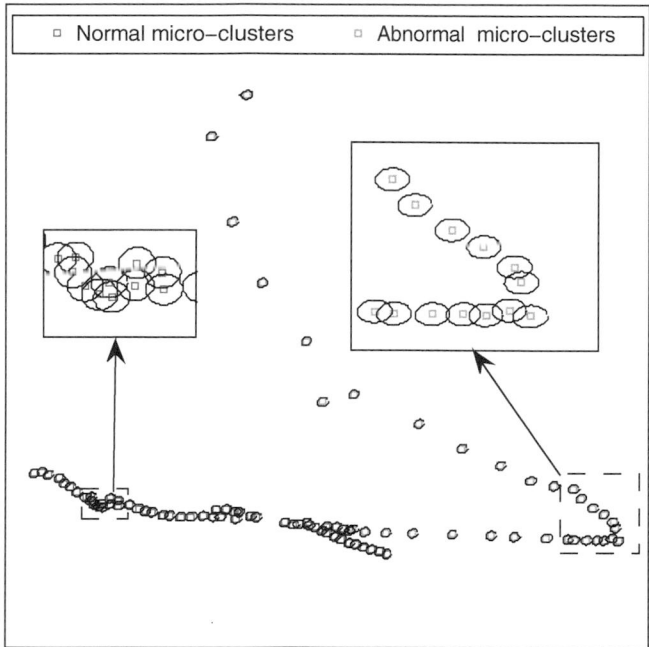

Figure 5.7 The extracted proximity-detection rules for two data points (attributes) in two-dimensional space and each rule is represented by a centroid of a microcluster.

because a large width will degrade the model accuracy, while the small width will result in many rules that will need to be checked in the detection phase. Since an inconsistency score for each observation is the average area of the neighborhood of that observation, the mean of inconsistency scores is proposed as an appropriate value of the fixed-width parameter γ and is defined as follows.

Let X be a vector of inconsistency scores of the training data set D_{points}, $X = \{x_1, x_2, ...x_n\}$, where n is the number of observations:

$$\gamma = \frac{1}{n} \sum_{i=1}^{n} X(i) \tag{5.4}$$

Figure 5.7 shows the proximity-detection rules for two data points (attributes), which in this case are the two sensors for humidity and temperature, where the consistent observations are clustered into microclusters using fixed-width clustering and the centroids of these produced microclusters are used as normal proximity-detection rules for monitoring the current system. Also, the inconsistent observations are clustered into microclusters and their respective centroids are used as abnormal proximity-detection rules.

Algorithm 5 summarizes the steps that are followed to extract proximity-based detection rules for both consistent/inconsistent behaviors.

Algorithm 5   The extraction algorithm of proximity-based detection rules

---

1 **Input:** w

   `/* Width of a cluster as eq 5.4`   `*/`

2 **Input:** data

   `/* List of observations for one behaviour whether`

      `consistent or inconsistent`   `*/`

3 Initialize the set of cluster $\xi \longleftarrow \emptyset$;

4 Initialize the number of clusters $M \longleftarrow 0$;

5 **foreach** *Training sample* $c_i \in Data$ **do**

6   **if** $M == 0$ **then**

7   Make a new cluster χ_1 with centroid χ_1' from c_i;

8       $\chi_1 \leftarrow \{c_i\}; \chi_1' \leftarrow c_i$;

9       $\xi \leftarrow \{\chi_1\}; M \leftarrow M + 1$;;

10   **else**

11   Find the closest cluster χ_n to c_i;

12       $n = \min_{i \in k} \Omega(c_i, \chi_n)$`/*` where $k = 1 \ldots M$ and Ω is the function of Euclidean distance   `*/`

13   **if** *The distance n* $< w$ **then**

14   Add c_i to cluster χ_n and update cluster centroid χ_n';

15          $\chi_n \leftarrow \chi_n \cup \{c_i\}$;

16   **else**

17          $M \leftarrow M + 1$;

18   Make a new cluster χ_M with centroid χ_M' from c_i;

19          $\chi_M \leftarrow \{c_i\}; \chi_M' \leftarrow c_i$;

20          $\xi \leftarrow \xi \cup \{\chi_M\}$;

21 **return** [All Clusters' centroids χ']`/*` The returned centroids will be used as proximity detection rules   `*/`

22 ;

5.2.4 Inconsistency Detection

The proximity-detection rules, whereby each rule is represented by a cluster centroid, are used to monitor any observation for the target system to judge whether the current observation is consistent or inconsistent. Therefore, a current observation is labeled with the label of the closest cluster. Let s_j be the current observation, $C = \{c_1, c_2, \ldots, c_n\}$ be the centroids of consistent clusters, and $I = \{i_1, i_2, \ldots, i_n\}$ be the centroids of inconsistent clusters. The closest consistent and inconsistent clusters to s_j are determined by the following equations, respectively:

$$c_{min} = \min_{c \in C} \Omega(s_j, c) \tag{5.5}$$

$$i_{min} = \min_{i \in I} \Omega(s_j, i) \qquad (5.6)$$

where Ω is Equation 5.1 that computes the Euclidean distance between two observations. In this case, the Euclidean distance between the current (testing) observation s_i and all microclusters, which represent the consistent behaviors, are calculated. Then, the smallest distance is set to the variable c_{min}. Similarly, the Euclidean distance between s_i and all microclusters that represent the inconsistent behaviours is computed and the smallest distance is set to the variable o_{min}. Afterwards, as defined in the following, the status of the current observation s_i is judged as consistent when the Euclidean distance in c_{min} is less than o_{min}. Otherwise, it is judged as inconsistent.

$$StateType(s_i) = \begin{cases} c_{min} < o_{min} & \text{Consistent} \\ Otherwise & \text{Inconsistent} \end{cases} \qquad (5.7)$$

5.3 EXPERIMENTAL SETUP

This section focuses on the setup of the experimental environment in order to evaluate the robustness of the proposed approach. In what follows, we describe two integrity attacks that target the simulation system. We also describe the data sets used and the experimental parameters chosen for this evaluation.

5.3.1 System Setup

As the proposed detection approach is intended to monitor SCADA systems using the values of data points, a supervised infrastructure (e.g., power energy grids or water supply network) needs to be built to evaluate the approach. Chapter 3 described SCADAVT as a tesbed that could be used to build and evaluate SCADA applications on a large scale. Here SCAVAT is used to simulate a water distribution system (WDS) as the supervised infrastructure.

To simulate the supervised infrastructure, the library of the well-known and free hydraulic and water quality model, called EPANET (Lewis 1999), is used to design a WDS server to act as a surrogate for a real WDS. The EPANET model involves three modules, namely hydraulic, water quality, and water consumption modules. We fed the consumption module with a specific model (i.e., the 2010 Melbourne water consumption; Melbourne Water, 2012) so as to simulate the realistic behavior of a water distribution system. One virtual machine is assigned to the WDS server. This server feeds the simulated data to virtualized field devices, and receives the Modbus/TCP control messages via a proxy. This proxy is used as an interface between virtualized field devices and the WDS server. For realistic simulation, the WDS server reads and controls process parameters such as water flow, pressure, and valve status in response to message commands from a field device. The manipulated process parameters in the WDS server are as follows:

- Water flow, pressure, demand, and level
- Valve status and setting
- Pump status and speed

5.3.2 WDS Scenario

Figure 5.8 depicts an example of a simple WDS for a small town. This type of town is divided into three areas, namely A, B, and C. Each area has an elevated tank to supply it with water at a satisfactory pressure level. The supplied water is pumped out by three pumps from the treatment system into $Tank_1$. The water is also delivered to $Tank_2$ by two pumps. $Tank_3$ is supplied through gravity because of the elevation of $Tank_2$, which is higher than $Tank_3$. $Tank_1$ is twice as big as $Tank_2$ and $Tank_3$ because it is considered to be the main water source for areas B and C.

The water network is monitored and controlled by the SCADA system. In this scenario, four $RTUs$, namely RTU_1 … RTU_4, are used. The MUT server plays a key role in sending command messages, in addition to storing the acquired data in the

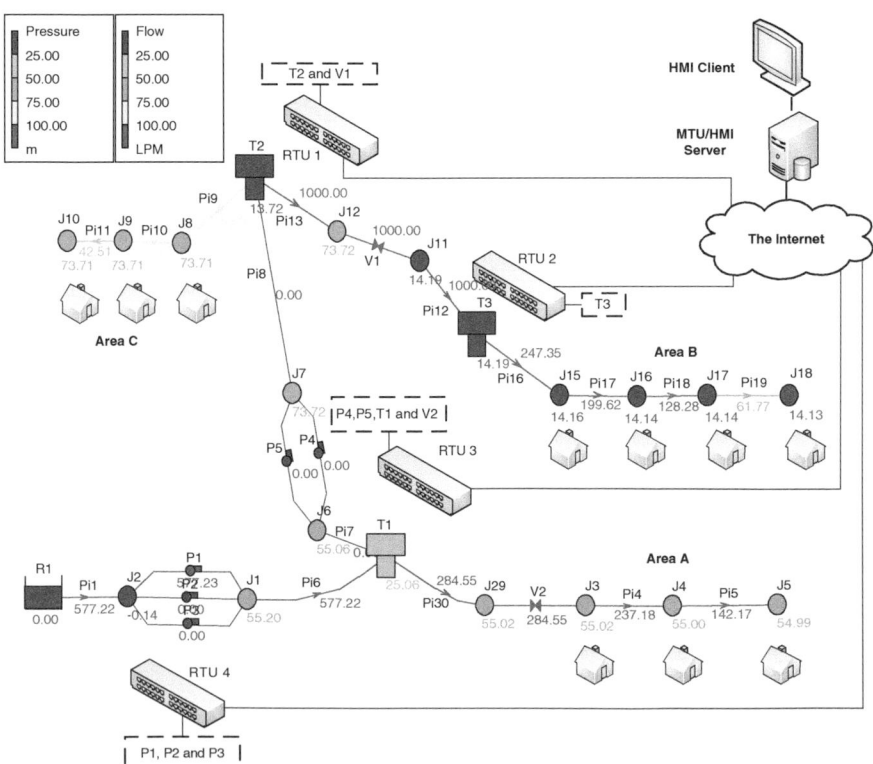

Figure 5.8 Simulation of a water distribution system.

Historian. The *MUT* server sends command messages to the *RTU* s to functionally perform as follows:

- RTU_4 to control status and speed of the pumps P_1, P_2, and P_3 according to the water level reading of *Tank*$_1$.
- RTU_3 to control status and speed of the pumps P_4 and P_5 according to the water level reading of *Tank*$_2$, and to read the water level of *Tank*$_1$.
- RTU_1 to open and close the valve V_1 according to the water level reading of *Tank*$_3$, and to read the water level of *Tank*$_2$.
- RTU_2 to read the water level of *Tank*$_3$.

5.3.3 Attack Scenario

The purpose of this scenario is to affect WDS's normal behavior. The public network (e.g., Internet) is used to interconnect all the WDS's components through Ethernet modules. The Modbus/TCP application protocol is set up as a communication protocol. However, all TCP vulnerabilities are inherited and therefore the system is susceptible to external attacks such as DoS (Denial of Service) attacks and spoofing.

We have opted to simulate *man-in-the-middle attacks*. These types of attacks require a prior knowledge of the target system and this can be obtained by the specifications or by the correlation analysis for the network traffic of that system. As mentioned from the specifications of the simulation system, RTU_4 is used to control the status and speed of pumps P_1, P_2, and P_3 in accordance with the water level reading of *Tank*$_1$. This is automatically performed by the MUT server. We compromised the MUT server and performed two malicious actions as follows.

Firstly, we sent 100 successive command messages to the RTU_4 to turn off *pump*$_2$ and *pump*$_3$, and randomly changed the the speeds of *pump*$_1$ between 0.1 and 1. Secondly, we performed a similar malicious action on RTU_4 to turn off *pump*$_2$ and randomly change the speeds of *pump*$_1$ and *pump*$_2$ between 0.1 to 1. These types of attacks can be launched in a number of ways and RTU_4 will trust such commands as they are sent from the MUT server.

5.3.4 Data Sets

To provide quantitative results for the proposed SDAD approach, we use seven labeled data sets: five are publicly available (Suthaharan et al. 2010; Frank and Asuncion, 2013b) and two are generated by the simulation system.

Simulated Data Sets. Since SDAD is mainly intended for SCADA systems using the values of data points (process parameters), the Water Distribution System (WDS) that has been simulated using the *SCADAVT* to generate data sets for the evaluation. As previously discussed in Chapter 3, the *MTU* server plays a key role in sending command messages to the PLCs for controlling and monitoring purposes. To create the data sets, all communication messages between *MTU* and all PLCs are collected under normal/abnormal conditions. All SCADA data $X = \{x_1, x_2, \dots, x_n\}$ that are observed at a period of time t are represented by a vector $v_t = \{x_{1,t}, x_{2,t}, \dots, x_{n,t}\}$. The simulated data vectors $V = \{v_1, v_2, \dots, v_m\}$, which are observed under normal conditions, are labeled as normal, while vectors $V = \{v_{m+1}, v_{m+2}, \dots, v_{m+n}\}$, which are observed under abnormal conditions, are labeled as abnormal.

Under the abnormal conditions, *man-in-the-middle attacks* are simulated. Since we had prior knowledge of the target system, integrity attacks that violated the specifications of the system are launched. This was performed by compromising the MTU server and two malicious actions were performed as follows. Firstly, 100 successive command messages are sent to PLC4 to turn off *pump₂* and *pump₃* and randomly change the speed of *pump₁* between 0.1 to 1. Secondly, a similar malicious attack is launched on PLC4 to turn off *pump₂* and randomly change the speed of *pump₁* and *pump₂* between 0.1 to 1. These types of attacks can be launched in a number of ways, and PLC4 will trust such commands as they are sent from the MTU server.

Two simulated data sets are generated, each consisting of 113 data points (attributes) and approximately more than 10,000 instances. In addition, each data set has 100 inconsistent observations, where each inconsistent observation in each data set is generated by a different attack. The simulated data sets will be denoted as *SimData*1 and *SimData*2.

Real Data Sets.   For more quantitative results, SDAD is also evaluated here using five real data sets. One of them comes from the daily measures of sensors in an urban waste water treatment plant (referred to as *DUWWTP*), and consists of 38 data points (attributes) (Frank and Asuncion, 2013b). This data set consists of approximately 527 instances, while 14 instances are labeled as abnormal. The four other data sets are collected from a real wireless sensor network (Suthaharan et al., 2010), and each consists of two attributes (temperature and humidity). Each data set has more than 4000 instances, and a tiny portion of abnormal instances. For simplicity, we refer to these data sets, Multi-hop Outdoor Real Data, Multi-hop Indoor Real Data, Single-hop Outdoor Real Data, and Single-hop Indoor Real Data, as *MORD*, *MIRD*, *SORD*, and *SIRD*.

5.3.5   Normalization

To improve the accuracy and efficiency of SDAD, the normalization technique is applied to all testing data sets to scale features by a range 0.0 of 1.0. This will prevent features with a large scale from outweighing features with a small scale. As the actual minimum/maximum of features are already known, and also because the identification process is performed in static mode, a min-max normalization technique is used to map the values of features. A given feature A will have values in [0.0, 1.0]. Let us denote by min_A and max_A the minimum value and maximum value of A respectively.

Then, to produce the normalized value of v ($v \in A$) using the min-max normalization method, which we denote as $\acute{v}$, the following formula is used:

$$\acute{v} = \frac{v - min_A}{max_A - min_A} \tag{5.8}$$

Two parameters are required for SDAD:

- The k-nearest neighbor parameter is the influencing factor for the anomaly scoring technique. However, this value is insensitive and can be heuristically determined based on the assumption that anomalies constitute a tiny portion of the data. Therefore, the value of k-nearest is set to be 1% of the representative data set, because this value is assumed to discriminate between abnormal instances and normal ones in terms of the density-based distance.
- The cut-off threshold η is used to select the most relevant anomalies after applying the inconsistency scoring technique. In this experiment, 10 values (from 0.50% to 5% increased by 0.50%) of η are tested and the best values are conditioned by the high detection rate of inconsistent observations and low detection of the false positive rate. Therefore, any value of η that yields a detection rate more than or equal to 90% with a false positive rate less than or equal to 1.50% is considered as a promising threshold to efficiently separate inconsistent observations from consistent ones and, consequently, the robust proximity-detection rules can be extracted from these.

5.4 RESULTS AND ANALYSIS

This section aims to evaluate the accuracy of SDAD. Two parts in this approach are evaluated: the first part is the accuracy of the separation between consistent and inconsistent observations as a first phase to extract detection rules, while the second part is the detection accuracy of these extracted detection rules. The performance of the second part is compared with the k-means algorithm (MacQueen et al., 1967) that is considered as one of the most useful and promising methods that can be adapted to build an unsupervised clustering-based intrusion detection model (Jianliang et al. 2009; Münz et al., 2007). All observations in the testing data sets, as discussed earlier, are used to demonstrate the accuracy of the process that separates consistent from inconsistent observations in each data set. To evaluate the robustness of the extracted detection rules in detecting inconsistent observations, k fold cross-validation is applied to each data set for each experimental parameter. The k-foldcross-validation method (Hastie et al., 2009) is performed by dividing the data set into k equal sized subsets. Each time one of the subsets is used as the validation data to test the model, while the remaining k-1 subsets are used to train the model. In this evaluation, k is set to 10, as suggested by Kohavi (1995), in order to reliably demonstrate the appropriateness of the proposed predictive model.

5.4.1 Accuracy Metrics

Several metrics have been used to evaluate anomaly detection approaches. In this evaluation, the *precision*, *recall* (detection rate), *false positive rate*, and *F-Measure* metrics are used to quantitatively measure the performance of the system. The first phase of the proposed approach, which is responsible for separating inconsistent observations from consistent ones, is evaluated using *Detection rate* and *false positive rate* metrics. This is because the separation of most "true" inconsistent observations from consistent ones with less false positive rate will yield robust proximity-detection rules for both inconsistent/consistent observations. *F-Measure* metric, however, is used to evaluate the efficiency of the extracted proximity-detection rules as it is not dependent on the size of the training and testing data set. Further details about these metrics are as follows:

$$Recall\ (Detection\ rate) = \frac{TP}{TP + FN} \tag{5.9}$$

$$False\ positive\ rate = \frac{FP}{FP + TN} \tag{5.10}$$

$$Precision = \frac{TP}{TP + FP} \tag{5.11}$$

$$F - Measure = 2 \times \frac{precision \times recall}{precision + recall} \tag{5.12}$$

where *TP* is the number of inconsistent observations that are correctly detected, *FN* is the number of inconsistent observations that have occurred but have not been detected, *FP* are the consistent observations that have been incorrectly flagged as inconsistent, and *TN* is the number of consistent observations that have been correctly classified. The recall (detection rate) is the proportion of correctly detected inconsistent observations to the actual size of the inconsistent observations in the testing data set, while the false positive rate is the proportion of incorrectly classified consistent observations as inconsistent to the actual size of the consistent observations in the testing data set. The precision metric is used to demonstrate the robustness of the IDS in minimizing the false positive rate. However, the system can obtain a high precision score although a number of inconsistent observations have been missed. Similarly, the system can obtain a high recall score although the false positive rate is higher. Therefore, the F-Measure, which is the harmonic mean of precision and recall, would be a more appropriate metric for demonstrating the accuracy of IDS approaches.

5.4.2 Separation Accuracy of Inconsistent Observations

As shown in Figure 5.3, the separation of inconsistent observations from the unlabeled training data set should be performed prior to the extraction of these proximity-detection rules. Therefore, it is important to evaluate the accuracy of the separation process of the most "true" inconsistent observations. This is because proximity detection rules for both normal and abnormal behaviors of the target system are extracted from the consistent and inconsistent observations, respectively.

From the intensive investigation, the best extracted proximity detection rules that represent the normal behavior of a system are the ones extracted from actual normal (consistent) observations, and this is true with proximity detection rules that represent the abnormal behavior. Therefore, the performance of the separation phase is measured by two metrics: the high detection rate and the false positive rate. This is because the separation of the most "true" inconsistent observations with a very small false positive rate produces the two most pure data sets of consistent/inconsistent observations. It is important to determine the near-optimal value of the separation threshold η. Since we assign an inconsistency score for each observation, the inconsistent ones are assumed to have the highest scores. Therefore, we use a range of cut-off thresholds, where each threshold η represents the top percentage of the observations that have the highest inconsistency scores. Based on the empirical results, the near-optimal threshold η, which can separate inconsistent observations from consistent ones to extract robust proximity-detection rules, is the one that can meet the following criteria:

$$\begin{cases} Detection\ rate \geq 90\% \\ False\ positive\ rate \leq 1.5\% \end{cases} \tag{5.13}$$

Tables 5.1 to 5.7 show the separation accuracy of inconsistent observations in all testing data sets using a range of cut-off thresholds η. It can be seen that the significant results, which meet the criterion in Equation 5.1, are shown in a bold font style. Clearly, the use of an inconsistency scoring method, which was introduced in Section 5.2.2, demonstrated significant results in separating inconsistent observations from the unlabeled training data set. That is, the most inconsistent observations can be separated from consistent ones. However, there is no fixed threshold that can work with all data sets and this can be attributed to a number of characteristics for each data set, such as the distribution, the number of inconsistent observations, and the application domain for each data set.

5.4.3   Detection Accuracy

The aim of this evaluation is to illustrate the detection accuracy of the proposed SDAD approach and the k-means algorithm (MacQueen et al., 1967) that is proposed as a useful and promising method for building an unsupervised clustering-based intrusion detection model (Jianliang et al., 2009; Münz et al., 2007). Since the k-means algorithm that we have chosen as a basis for comparison is inherently different in terms of the required parameters for building a clustering-based intrusion detection model, we tested it separately from the proposed model. The 10-fold validation method is used to evaluate the generalization detection accuracy of both SDAD and k-means. Each data set is divided equally into 10 subsets where, in each fold, a different subset is used as a testing data set, while the remaining nine subsets are used as a training data set.

TABLE 5.1   The Separation Accuracy of Inconsistent Observations on DUWWTP

η	Detection rate	False positive	Precision	F-Measure
0.50%	14.29%	0.00%	100.00%	25.00%
1.00%	35.71%	0.00%	100.00%	52.63%
1.50%	50.00%	0.00%	100.00%	66.67%
2.00%	70.71%	0.02%	99.00%	82.50%
2.50%	79.29%	0.19%	92.50%	85.38%
3.00%	87.14%	0.39%	87.14%	87.14%
3.50%	**92.86%**	**0.87%**	76.47%	83.87%
4.00%	**92.86%**	**1.30%**	68.42%	78.79%
4.50%	94.29%	1.69%	62.86%	75.43%
5.00%	100.00%	2.17%	58.33%	73.68%

η: Top percentage of observations, which are sorted by inconsistency scores in ascending order, are assumed as inconsistent.

TABLE 5.2   The Separation Accuracy of Inconsistent Observations on SimData1

η	Detection rate	False positive	Precision	F-Measure
0.50%	40.78%	0.06%	88.51%	55.84%
1.00%	80.78%	0.13%	86.74%	83.65%
1.50%	**98.04%**	**0.45%**	70.42%	81.97%
2.00%	**98.04%**	**0.95%**	52.91%	68.73%
2.50%	**98.04%**	**1.46%**	42.19%	59.00%
3.00%	98.04%	1.97%	35.21%	51.81%
3.50%	98.04%	2.47%	30.21%	46.19%
4.00%	98.04%	2.97%	26.46%	41.67%
4.50%	98.04%	3.48%	23.47%	37.88%
5.00%	98.04%	3.99%	21.14%	34.78%

TABLE 5.3   The Separation Accuracy of Inconsistent Observations on SimData2

η	Detection rate	False positive	Precision	F-Measure
0.50%	46.50%	0.01%	98.94%	63.27%
1.00%	81.50%	0.14%	85.79%	83.59%
1.50%	**100.00%**	**0.45%**	70.42%	82.64%
2.00%	**100.00%**	**0.95%**	52.91%	69.20%
2.50%	**100.00%**	**1.46%**	42.19%	59.35%
3.00%	100.00%	1.97%	35.21%	52.08%
3.50%	100.00%	2.47%	30.21%	46.40%
4.00%	100.00%	2.97%	26.46%	41.84%
4.50%	100.00%	3.48%	23.47%	38.02%
5.00%	100.00%	3.98%	21.14%	34.90%

TABLE 5.4   The Separation Accuracy of Inconsistent Observations on SIRD

η	Detection rate	False positive	Precision	F-Measure
0.50%	17.09%	0.00%	100.00%	29.20%
1.00%	34.19%	0.00%	100.00%	50.96%
1.50%	51.28%	0.00%	100.00%	67.80%
2.00%	68.38%	0.00%	100.00%	81.22%
2.50%	85.47%	0.00%	100.00%	92.17%
3.00%	**100.00%**	**0.08%**	97.50%	98.73%
3.50%	**100.00%**	**0.59%**	83.57%	91.05%
4.00%	**100.00%**	**1.09%**	73.58%	84.78%
4.50%	100.00%	1.60%	65.36%	79.05%
5.00%	100.00%	2.12%	58.79%	74.05%

TABLE 5.5   The Separation Accuracy of Inconsistent Observations on SORD

η	Detection rate	False positive	Precision	F-Measure
0.50%	71.88%	0.00%	100.00%	83.64%
1.00%	**90.94%**	**0.35%**	64.67%	75.58%
1.50%	**93.75%**	**0.84%**	44.12%	60.00%
2.00%	93.75%	1.77%	27.27%	42.25%
2.50%	93.75%	2.00%	25.00%	39.47%
3.00%	96.88%	2.33%	22.79%	36.90%
3.50%	96.88%	2.84%	19.50%	32.46%
4.00%	96.88%	3.35%	17.03%	28.97%
4.50%	96.88%	3.84%	15.20%	26.27%
5.00%	96.88%	4.35%	13.66%	23.94%

TABLE 5.6   The Separation Accuracy of Inconsistent Observations on MORD

η	Detection rate	False positive	Precision	F-Measure
1.00%	72.41%	0.00%	100.00%	84.00%
1.50%	86.03%	0.31%	79.21%	82.48%
2.00%	86.21%	0.84%	58.82%	69.93%
2.50%	86.55%	1.34%	47.36%	61.22%
3.00%	87.41%	1.83%	39.92%	54.81%
3.50%	87.93%	2.33%	34.46%	49.51%
4.00%	87.93%	2.83%	30.18%	44.93%
4.50%	88.45%	3.33%	27.00%	41.37%
5.00%	89.31%	3.82%	24.55%	38.51%

TABLE 5.7   The Separation Accuracy of Inconsistent Observations on MIRD

η	Detection rate	False positive	Precision	F-Measure
0.50%	21.00%	0.00%	100.00%	34.71%
1.00%	42.00%	0.00%	100.00%	59.15%
1.50%	63.00%	0.00%	100.00%	77.30%
2.00%	85.00%	0.00%	100.00%	91.89%
2.50%	**100.00%**	**0.15%**	94.34%	97.09%
3.00%	**100.00%**	**0.65%**	78.74%	88.11%
3.50%	**100.00%**	**1.16%**	67.57%	80.65%
4.00%	100.00%	1.67%	59.17%	74.35%
4.50%	100.00%	2.18%	52.63%	68.97%
5.00%	100.00%	2.71%	47.17%	64.10%

In this evaluation, the optimal detection model is the one that can detect all inconsistent observations and has a zero false positive rate. In some cases, a detection model can have a zero false positive rate, but a number of inconsistent observations are missed, while in another case, the detection model can have a high false positive rate, but detects all inconsistent observations. Therefore, the F-Measure, which is the harmonic mean of precision and recall (detection rate), would be a more appropriate metric to demonstrate the detection accuracy. We assumed that F-measure results that are greater than or equal to 90 are significant and they appear in bold font.

k-Means Algorithm. k-Means (MacQueen et al., 1967) is a clustering algorithm that groups similar data into a number of clusters based on a specified number of clusters k, where the value of k is specified in advance. To build the detection model using the k-means algorithm, the unlabeled training data set is clustered into a number of clusters, and each produced cluster is labeled as either normal or abnormal. The labeling method is based on the percentage of data in each cluster. For instance, if the members of a cluster represent the percentage θ of the overall training data, it is labeled as an abnormal cluster; otherwise it is labeled as normal. A range of numbers of clusters from 10 to 150 increased by 10 are tested in this evaluation, and we have chosen only the best six values, 50, 60, 70, 80, 90, 100, and with each value we have tested five values of θ, which are 0.01, 0.02, 0.03, 0.04, 0.05. The selection of the values of θ is based on the two assumptions about the assumed size and the characteristics of anomalies in unlabeled data (Portnoy et al., 2001). For instance, given θ is set to 0.01, if any produced cluster whose members are less or equal to 0.01% of the overall members in the training data set, it is labeled as abnormal; otherwise, it is labeled as normal. In the testing phase, the distance between the centroids of the produced clusters and each observation in the testing data set is calculated and is given the label of its closest cluster.

Tables 5.8 to 5.14 show the accuracy results in detecting inconsistent observations in all used data sets, and it can be seen that the best numbers of clusters are relatively large. This is because the smaller the value of the numbers of clusters, the higher is

TABLE 5.8   The Detection Accuracy Results of k-Means in Detecting Consistent/Inconsistent Observations on DUWWTP

#Clusters	θ	Detection rate	False positive	Precision	F-Measure
50	1%	62.86%	1.56%	84.57%	69.65%
	2%	80.71%	14.22%	59.25%	59.98%
	3%	87.14%	29.74%	45.76%	50.50%
	4%	90.36%	42.57%	37.97%	44.22%
	5%	92.29%	53.01%	32.90%	39.86%
60	1%	67.14%	6.24%	61.00%	62.36%
	2%	83.57%	21.16%	44.96%	53.36%
	3%	89.05%	39.70%	35.04%	44.37%
	4%	91.79%	52.77%	29.53%	39.03%
	5%	93.43%	61.62%	26.10%	35.62%
70	1%	70.00%	6.63%	58.82%	62.95%
	2%	84.29%	28.86%	39.88%	48.73%
	3%	89.52%	47.82%	31.18%	40.56%
	4%	92.14%	59.45%	26.56%	36.05%
	5%	93.71%	67.29%	23.68%	33.17%
80	1%	75.71%	12.10%	46.84%	57.08%
	2%	87.86%	36.27%	32.71%	44.19%
	3%	91.91%	54.20%	26.20%	37.23%
	4%	93.93%	65.65%	22.65%	33.28%
	5%	95.14%	72.52%	20.52%	30.91%
90	1%	88.57%	16.58%	45.11%	58.62%
	2%	94.29%	45.73%	30.31%	42.73%
	3%	96.19%	62.53%	24.36%	35.87%
	4%	97.14%	71.90%	21.27%	32.26%
	5%	97.71%	77.52%	19.42%	30.10%
100	1%	91.43%	22.99%	36.21%	51.52%
	2%	95.71%	51.74%	25.46%	38.56%
	3%	97.14%	66.91%	21.08%	33.01%
	4%	97.86%	75.14%	18.82%	30.13%
	5%	98.29%	80.11%	17.45%	28.39%

θ: The percentage of the data in a cluster to be assumed as malicious.

the expectation of outlying observations, which are assumed to be inconsistent, to be grouped into the large normal clusters. Clearly, the use of k-means to build the detection model for detecting inconsistent observations demonstrated poor detection results. As can be seen, we can obtain a high detection rate using k-*means*; however, it is obvious that the false positive rate is very high and this can be a nuisance for a security administrator by warning him with false positive alarms.

SDAD Evaluation.   This section evaluates the detection accuracy of the SDAD approach. As previously discussed, SDAD in the first phase separates the inconsistent observations from the consistent ones in the unlabeled training data set, and

TABLE 5.9   The Detection Accuracy Results of *k*-Means in Detecting Consistent/Inconsistent Observations on SimData1

#Clusters	θ	Detection rate	False positive	Precision	F-Measure
50	1%	67.65%	3.63%	41.29%	50.56%
	2%	60.98%	21.06%	23.73%	30.82%
	3%	67.39%	38.04%	17.52%	23.74%
	4%	70.64%	51.21%	14.18%	19.79%
	5%	76.47%	60.34%	12.31%	17.67%
60	1%	68.63%	8.05%	25.59%	37.04%
	2%	67.84%	29.08%	15.88%	24.17%
	3%	73.14%	47.39%	12.14%	19.06%
	4%	67.75%	59.33%	9.75%	15.51%
	5%	74.20%	67.05%	8.75%	14.23%
70	1%	69.02%	14.36%	16.88%	26.97%
	2%	76.28%	42.18%	11.20%	18.66%
	3%	77.71%	58.84%	8.82%	15.02%
	4%	80.88%	68.68%	7.69%	13.32%
	5%	84.71%	74.94%	7.09%	12.44%
80	1%	97.26%	21.60%	18.24%	30.68%
	2%	90.69%	50.17%	11.64%	20.10%
	3%	91.44%	65.26%	9.28%	16.30%
	4%	93.58%	73.40%	8.16%	14.51%
	5%	94.86%	78.72%	7.46%	13.39%
90	1%	89.02%	25.62%	14.30%	24.61%
	2%	94.51%	57.33%	9.77%	17.29%
	3%	96.34%	71.55%	8.07%	14.50%
	4%	92.40%	78.23%	7.01%	12.69%
	5%	93.92%	82.58%	6.54%	11.94%
100	1%	98.63%	36.30%	11.85%	21.15%
	2%	99.31%	64.75%	8.43%	15.34%
	3%	99.54%	76.38%	7.18%	13.22%
	4%	97.26%	81.90%	6.46%	11.96%
	5%	97.80%	85.52%	6.10%	11.35%

a range of different separation thresholds η are used in this process. As a result, both sets of inconsistent and consistent observations are expected to be different for each threshold η. Therefore, proximity-detection rules are independently extracted for each different threshold η and evaluated in detecting inconsistent observations. Since the fixed-width clustering algorithm is adapted to individually cluster the inconsistent and consistent observations into microclusters with constant width, the value of the width parameter w is required. However, the value of w can be obtained using Equation 5.4. The microclusters created from the consistent observations are labeled as normal, while the microclusters created from inconsistent observations are labeled as abnormal. Figure 5.7 illustrates how proximity-detection rules for two data points (attributes) are extracted.

TABLE 5.10   **The Detection Accuracy Results of k-Means in Detecting Consistent/Inconsistent Observations on SimData2**

#Clusters	θ	Detection rate	False positive	Precision	F-Measure
50	1%	100.00%	3.39%	60.45%	74.79%
	2%	100.00%	19.02%	36.41%	48.40%
	3%	100.00%	36.54%	26.39%	36.25%
	4%	100.00%	50.22%	21.05%	29.58%
	5%	100.00%	59.38%	17.80%	25.49%
60	1%	100.00%	8.96%	35.90%	52.54%
	2%	100.00%	31.06%	22.11%	33.94%
	3%	100.00%	48.85%	16.54%	26.06%
	4%	100.00%	61.23%	13.57%	21.77%
	5%	100.00%	68.98%	11.78%	19.17%
70	1%	100.00%	12.49%	28.19%	43.87%
	2%	100.00%	38.28%	17.60%	28.49%
	3%	100.00%	56.73%	13.37%	22.10%
	4%	100.00%	67.18%	11.19%	18.80%
	5%	100.00%	73.75%	9.87%	16.80%
80	1%	100.00%	19.61%	19.93%	33.18%
	2%	100.00%	49.31%	12.85%	22.04%
	3%	100.00%	64.58%	10.17%	17.76%
	4%	100.00%	73.07%	8.79%	15.55%
	5%	100.00%	78.46%	7.95%	14.19%
90	1%	100.00%	28.81%	14.46%	25.24%
	2%	100.00%	56.83%	9.92%	17.73%
	3%	100.00%	70.25%	8.19%	14.83%
	4%	100.00%	77.44%	7.30%	13.33%
	5%	100.00%	81.96%	6.76%	12.42%
100	1%	100.00%	34.32%	12.39%	22.03%
	2%	100.00%	62.78%	8.70%	15.79%
	3%	100.00%	74.48%	7.36%	13.51%
	4%	100.00%	80.86%	6.67%	12.33%
	5%	100.00%	84.69%	6.25%	11.62%

Tables 5.15 to 5.21 show the detection accuracy of SDAD for all used data sets. Each table demonstrates the detection accuracy results for a particular data set, where each row of the table represents detection accuracy results of the extracted proximity-detection rules for a specific threshold η. For instance, the first row in Table 5.15 indicates that the separation threshold η is set to 0.50% and the cluster width parameter is set to 0.7506 to create microclusters for both consistent/inconsistent observations. In this example, the identified consistent and inconsistent observations are partitioned into 62 and 2 microclusters, respectively. The centroids of these microclusters are used as proximity-detection rules for monitoring inconsistent observations for a given system.

TABLE 5.11 The Detection Accuracy Results of k-Means in Detecting
Consistent/Inconsistent Observations on SIRD

#Clusters	θ	Detection rate	False positive	Precision	F-Measure
50	1%	95.25%	5.88%	69.20%	80.09%
	2%	97.12%	19.54%	49.40%	62.74%
	3%	97.74%	32.45%	39.26%	52.44%
	4%	98.18%	44.10%	33.14%	45.76%
	5%	98.54%	52.82%	29.24%	41.40%
60	1%	96.95%	10.21%	56.97%	71.64%
	2%	98.22%	29.33%	39.51%	53.86%
	3%	98.48%	42.83%	31.79%	45.25%
	4%	98.73%	54.88%	27.11%	39.72%
	5%	98.98%	62.54%	24.26%	36.34%
70	1%	97.63%	14.21%	48.66%	64.89%
	2%	98.31%	37.05%	33.63%	48.09%
	3%	98.64%	52.35%	27.15%	40.33%
	4%	98.98%	62.47%	23.59%	35.97%
	5%	99.09%	69.07%	21.38%	33.23%
80	1%	98.81%	21.23%	39.63%	56.39%
	2%	98.90%	42.42%	28.63%	43.16%
	3%	99.15%	58.49%	23.47%	36.51%
	4%	99.24%	67.72%	20.74%	32.95%
	5%	99.39%	73.85%	19.04%	30.72%
90	1%	98.98%	27.56%	33.09%	49.58%
	2%	99.49%	49.87%	24.55%	38.59%
	3%	99.55%	63.57%	20.74%	33.45%
	4%	99.66%	72.45%	18.59%	30.51%
	5%	99.73%	77.66%	17.32%	28.77%
100	1%	99.66%	37.35%	26.92%	42.35%
	2%	99.66%	57.57%	20.95%	34.20%
	3%	99.77%	70.66%	18.11%	30.16%
	4%	99.83%	77.99%	16.60%	28.01%
	5%	99.86%	82.40%	15.69%	26.71%

As shown in Tables 5.15 to 5.21, SDAD has achieved significant results for all data sets except *MORD* (see Table 5.20), where its results are nearly significant. However, the significant results for each data set are achieved by using variant separation thresholds η. For instance, the significant results on the data set *DUWWTP*, which are shown in Table 5.15, are achieved when η is set at 3.50% and 4.00%, while on the data sets, *SimData1* and *SimData2*, which are shown in Tables 5.16 and 5.17 respectively, are achieved when η is set to 1.50%, 2.00%, and 2.50%. Tables 5.18 and 5.21 show the results of the detection accuracy of the data sets *SIRD* and *MIRD*, respectively, and it can be seen that the significant results are achieved when η is set to 3.00%, 3.50%, and 4.00% for *SIRD*, while η is set to 2.50%, 3.00%, and 3.50% for

TABLE 5.12   The Detection Accuracy Results of *k*-Means in Detecting Consistent/Inconsistent Observations on SORD

#Clusters	θ	Detection rate	False positive	Precision	F-Measure
50	1%	84.38%	6.99%	28.61%	42.53%
	2%	87.81%	21.17%	18.21%	28.45%
	3%	90.00%	35.70%	13.64%	21.83%
	4%	91.25%	48.44%	11.08%	18.01%
	5%	92.63%	57.79%	9.50%	15.64%
60	1%	90.00%	10.12%	24.06%	37.43%
	2%	92.19%	29.13%	15.01%	24.32%
	3%	94.17%	46.41%	11.26%	18.62%
	4%	95.16%	58.89%	9.23%	15.49%
	5%	96.00%	66.57%	8.02%	13.62%
70	1%	93.13%	16.49%	15.62%	26.69%
	2%	94.38%	39.01%	10.21%	17.91%
	3%	95.42%	55.33%	7.95%	14.15%
	4%	96.09%	65.31%	6.76%	12.16%
	5%	96.75%	72.01%	6.03%	10.93%
80	1%	95.00%	23.98%	11.46%	20.41%
	2%	95.94%	49.16%	7.73%	14.05%
	3%	97.08%	63.45%	6.27%	11.52%
	4%	97.34%	71.69%	5.49%	10.17%
	5%	97.75%	77.13%	5.01%	9.34%
90	1%	96.25%	30.27%	9.34%	17.01%
	2%	97.50%	55.19%	6.56%	12.15%
	3%	98.13%	69.30%	5.43%	10.14%
	4%	98.59%	76.98%	4.84%	9.11%
	5%	98.88%	81.58%	4.49%	8.49%
100	1%	96.25%	38.15%	7.60%	14.07%
	2%	98.13%	61.47%	5.62%	10.54%
	3%	97.92%	72.74%	4.80%	9.07%
	4%	98.44%	79.55%	4.37%	8.31%
	5%	98.63%	83.57%	4.12%	7.84%

MIRD. Interestingly, the significant results of the data set *SORD*, which are shown in Table 5.19, are achieved with only small values of η, 1.00%, and 1.50%.

5.5 SDAD LIMITATIONS

Although the SDAD method demonstrated significant results on most data sets, there is no fixed separation threshold η that can work with all data sets. Table 5.22 exhibits the interesting separation thresholds η, which were near-optimal to separate inconsistent from consistent observations in the separation stage of SDAD, for each data set.

TABLE 5.13   The Detection Accuracy Results of k-Means in Detecting Consistent/Inconsistent Observations on MORD

#Clusters	θ	Detection rate	False positive	Precision	F-Measure
50	1%	85.17%	3.13%	65.00%	72.61%
	2%	87.76%	20.09%	39.26%	48.03%
	3%	91.15%	37.06%	28.83%	36.94%
	4%	93.28%	47.86%	23.44%	31.08%
	5%	94.48%	56.93%	20.00%	27.23%
60	1%	84.83%	7.23%	43.03%	56.72%
	2%	88.45%	28.80%	26.68%	37.64%
	3%	92.30%	46.15%	20.20%	29.59%
	4%	93.79%	57.57%	16.73%	25.16%
	5%	95.03%	65.43%	14.60%	22.42%
70	1%	87.24%	13.21%	29.85%	44.26%
	2%	91.03%	37.96%	19.27%	30.09%
	3%	93.91%	54.34%	15.08%	24.24%
	4%	95.35%	65.36%	12.80%	20.99%
	5%	96.28%	72.05%	11.43%	19.04%
80	1%	91.03%	18.85%	23.33%	37.11%
	2%	93.45%	46.92%	15.37%	25.44%
	3%	95.63%	62.46%	12.34%	20.90%
	4%	96.72%	71.22%	10.77%	18.52%
	5%	97.38%	76.89%	9.80%	17.05%
90	1%	87.24%	26.02%	17.63%	29.29%
	2%	91.90%	53.27%	12.32%	21.18%
	3%	94.60%	67.52%	10.26%	17.97%
	4%	95.95%	75.55%	9.17%	16.27%
	5%	96.76%	80.30%	8.52%	15.26%
100	1%	89.66%	33.98%	14.22%	24.53%
	2%	93.28%	58.75%	10.50%	18.61%
	3%	95.52%	72.07%	8.99%	16.16%
	4%	96.64%	78.95%	8.22%	14.91%
	5%	97.31%	83.16%	7.76%	14.15%

It can be seen that the maximum number of data sets that agree on one unique threshold η is three. This is because the unsupervised mode, where the labeled data is not available, relies mainly on assumptions (discussed earlier in Section 5.2.2) to distinguish between normal and abnormal behaviors. Therefore, the effectiveness of unsupervised anomaly-detection approaches is sensitive to parameter choices, especially when the boundaries between normal and abnormal behaviors are not clearly distinguishable. For instance, the anomaly-scoring-based technique, which is adapted in the first phase of the proposed approach and one of the famous unsupervised anomaly detection methods (Chandola et al., 2009b) assigns an anomaly score for each observation and actual abnormal observations are assumed to have the highest

TABLE 5.14   The Detection Accuracy Results of k-Means in Detecting
Consistent/Inconsistent Observations on MIRD

#Clusters	θ	Detection rate	False positive	Precision	F-Measure
50	1%	99.60%	4.18%	73.02%	84.02%
	2%	99.80%	19.54%	48.48%	61.31%
	3%	99.87%	33.13%	37.43%	49.73%
	4%	99.90%	45.79%	30.96%	42.47%
	5%	99.92%	55.80%	26.81%	37.69%
60	1%	100.00%	7.28%	60.81%	75.38%
	2%	100.00%	27.06%	39.92%	53.65%
	3%	100.00%	43.83%	30.73%	43.10%
	4%	100.00%	56.71%	25.62%	36.99%
	5%	100.00%	65.23%	22.47%	33.19%
70	1%	100.00%	12.99%	46.75%	63.42%
	2%	100.00%	38.03%	30.77%	44.59%
	3%	100.00%	55.80%	24.08%	36.16%
	4%	100.00%	66.20%	20.58%	31.70%
	5%	100.00%	72.75%	18.45%	28.97%
80	1%	100.00%	19.48%	36.21%	53.07%
	2%	100.00%	45.58%	24.74%	38.25%
	3%	100.00%	61.67%	19.96%	31.78%
	4%	100.00%	71.01%	17.45%	28.35%
	5%	100.00%	76.81%	15.93%	26.25%
90	1%	99.80%	26.28%	29.64%	45.60%
	2%	99.80%	52.94%	20.85%	33.55%
	3%	99.80%	67.49%	17.28%	28.50%
	4%	99.85%	75.62%	15.41%	25.84%
	5%	99.88%	80.49%	14.29%	24.25%
100	1%	100.00%	36.25%	23.31%	37.75%
	2%	100.00%	62.58%	17.12%	28.72%
	3%	100.00%	74.56%	14.73%	25.19%
	4%	100.00%	80.81%	13.51%	23.38%
	5%	100.00%	84.65%	12.78%	22.28%

scores. The key problem is how to find a near-optimal cut-off threshold that minimizes
the false positive rate while maximizing the detection rate.

5.6 CONCLUSION

This chapter described an unsupervised SCADA Data-driven Anomaly Detection
approach (SDAD) that generates, from unlabeled SCADA data, proximity anomaly
detection rules based on the clustering method. SDAD initially separates the
inconsistent observations from consistent ones in unlabeled SCADA data and then

TABLE 5.15   The Detection Accuracy of the Proximity-Detection Rules on DUWWTP

η	w	NC	AC	Detection rate	False positive	Precision	F-Measure
0.50%	0.7506	62	2	14.29%	0.00%	100.00%	25.00%
1.00%		59	5	35.71%	0.00%	100.00%	52.63%
1.50%		57	7	50.00%	0.00%	100.00%	66.67%
2.00%		54	10	70.71%	0.00%	100.00%	82.85%
2.50%		52	12	79.29%	0.00%	100.00%	88.45%
3.00%		51	14	87.14%	5.85%	80.26%	83.56%
3.50%		48	16	92.86%	0.78%	97.01%	**94.89%**
4.00%		46	18	92.86%	0.97%	96.30%	**94.55%**
4.50%		44	20	94.29%	9.75%	72.53%	81.99%
5.00%		41	23	100.00%	11.70%	70.00%	82.35%

η: Top percentage of observations, which are sorted by inconsistency scores in ascending order, are assumed as inconsistent.
w: The cluster width parameter.
NC: The number of the produced normal clusters. AC: The number of the produced abnormal clusters.

TABLE 5.16   The Detection Accuracy of the Proximity-Detection Rules on SimData1

η	w	NC	AC	Detection rate	False positive	Precision	F-Measure
0.50%	0.1456	316	39	40.78%	0.06%	98.58%	57.70%
1.00%		286	68	80.78%	0.13%	98.33%	88.70%
1.50%		265	92	98.04%	0.46%	95.42%	**96.71%**
2.00%		252	104	98.04%	1.11%	89.69%	**93.68%**
2.50%		242	116	98.04%	1.73%	84.75%	**90.91%**
3.00%		238	131	98.04%	2.49%	79.43%	87.76%
3.50%		234	138	98.04%	3.08%	75.76%	85.47%
4.00%		225	144	98.04%	3.64%	72.52%	83.37%
4.50%		219	151	98.04%	4.24%	69.40%	81.27%
5.00%		211	157	98.04%	4.75%	66.93%	79.55%

TABLE 5.17   The Detection Accuracy of the Proximity-Detection Rules on SimData2

η	w	NC	AC	Detection rate	False positive	Precision	F-Measure
0.50%	0.1476	307	42	46.50%	0.01%	99.79%	63.44%
1.00%		279	71	70.00%	0.14%	97.90%	81.63%
1.50%		262	90	100.00%	0.51%	94.97%	**97.42%**
2.00%		255	102	100.00%	1.18%	89.05%	**94.21%**
2.50%		243	112	100.00%	1.72%	84.82%	**91.79%**
3.00%		239	120	100.00%	2.42%	79.87%	88.81%
3.50%		231	130	100.00%	3.05%	75.93%	86.32%
4.00%		228	139	100.00%	3.63%	72.57%	84.10%
4.50%		222	146	100.00%	4.19%	69.64%	82.10%
5.00%		216	147	100.00%	4.71%	67.11%	80.32%

TABLE 5.18 The Detection Accuracy of the Proximity-Detection Rules on SIRD

η	w	NC	AC	Detection rate	False positive	Precision	F-Measure
0.50%	0.0124	63	17	17.09%	0.00%	100.00%	29.20%
1.00%		45	35	34.19%	0.00%	100.00%	50.96%
1.50%		36	44	51.97%	0.00%	100.00%	68.39%
2.00%		31	49	67.61%	0.00%	100.00%	80.67%
2.50%		24	57	72.65%	0.00%	100.00%	84.16%
3.00%		17	63	100.00%	0.42%	98.48%	**99.24%**
3.50%		15	64	100.00%	0.77%	97.26%	**98.61%**
4.00%		15	64	100.00%	1.58%	94.51%	**97.18%**
4.50%		15	64	100.00%	6.98%	79.59%	88.64%
5.00%		15	65	100.00%	8.14%	76.97%	86.99%

TABLE 5.19 The Detection Accuracy of the Proximity-Detection Rules on SORD

η	w	NC	AC	Detection rate	False positive	Precision	F-Measure
0.50%	0.0152	73	20	71.88%	0.00%	100.00%	83.64%
1.00%		67	27	90.94%	0.28%	95.41%	**93.12%**
1.50%		64	32	93.75%	0.90%	86.96%	**90.23%**
2.00%		62	33	93.75%	1.32%	81.97%	87.46%
2.50%		61	35	93.75%	2.02%	74.81%	83.22%
3.00%		57	37	96.88%	2.38%	72.26%	82.78%
3.50%		55	41	96.88%	3.03%	67.10%	79.28%
4.00%		55	42	96.88%	3.83%	61.75%	75.43%
4.50%		54	41	96.88%	4.19%	59.62%	73.81%
5.00%		54	42	96.88%	4.59%	57.41%	72.09%

TABLE 5.20 The Detection Accuracy of the Proximity-Detection Rules on MORD

η	w	NC	AC	Detection rate	False positive	Precision	F-Measure
0.50%	0.0093	74	14	36.21%	0.00%	100.00%	53.16%
1.00%		55	34	72.59%	0.00%	100.00%	84.12%
1.50%		53	38	77.59%	0.56%	94.54%	85.23%
2.00%		53	39	86.21%	1.10%	90.74%	88.42%
2.50%		52	39	86.55%	1.68%	86.55%	86.55%
3.00%		50	40	87.41%	2.20%	83.25%	85.28%
3.50%		49	41	87.93%	2.72%	80.19%	83.88%
4.00%		48	42	87.93%	3.00%	78.58%	82.99%
4.50%		48	43	88.45%	3.39%	76.57%	82.08%
5.00%		46	44	89.31%	3.99%	73.68%	80.75%

TABLE 5.21   The Detection Accuracy of the Proximity-Detection Rules on MIRD

η	w	NC	AC	Detection rate	False positive	Precision	F-Measure
0.50%	0.0135	62	13	20.10%	0.00%	100.00%	33.47%
1.00%		44	31	42.00%	0.00%	100.00%	59.15%
1.50%		37	38	63.00%	0.00%	100.00%	77.30%
2.00%		26	49	70.00%	0.00%	100.00%	82.35%
2.50%		20	55	100.00%	0.20%	99.11%	**99.55%**
3.00%		18	57	100.00%	0.87%	96.15%	**98.04%**
3.50%		18	58	100.00%	1.29%	94.43%	**97.13%**
4.00%		18	59	100.00%	6.54%	76.92%	86.96%
4.50%		15	61	100.00%	7.63%	74.07%	85.11%
5.00%		16	65	100.00%	8.28%	72.46%	84.03%

TABLE 5.22   The Illustration of the Acceptable Thresholds η That Produce Significant Accuracy Results of the Separation of Inconsistent Observations on Each Data Set

	Data sets							
η	DUWWTP	SimData1	SimData2	MORD	MIRD	SORD	SIRD	AgreeNO
0.50%								0
1.00%					√			1
1.50%		√	√		√			3
2.00%		√	√					2
2.50%		√	√				√	3
3.00%				√			√	2
3.50%	√			√			√	3
4.00%	√			√				2
4.50%								0
5.00%								0

η: Top percentage of observations, that are sorted by inconsistency scores in ascending order, are assumed as inconsistent. AgreeNO: The number of data sets that agree on each separation threshold η, where the agreement is judged by the significant detection results.

the proximity detection rules for each behaviour, whether consistent or inconsistent, are automatically extracted from the observations that belong to that behavior. The extracted proximity rules are used to monitor the abnormal behavior of SCADA systems by observing the evolution of SCADA data that are produced by the target SCADA data points. Experimental results of SDAD demonstrate its ability to separate inconsistent observations from consistent ones. Moreover, the automatically extracted, proximity-based detection rules show a significant detection accuracy rate compared with existing clustering-based intrusion-detection algorithms.

CHAPTER 6

A Global Anomaly Threshold to Unsupervised Detection

The effectiveness of unsupervised anomaly detection methods is sensitive to parameter choices, especially when the boundaries between normal and abnormal behavior are not clearly distinguishable. Therefore, the described SDAD method in Chapter 5 is based on an assumption by which anomalies is defined, where this assumption are controlled by a parameter choice. This chapter describe a new method, called GATUD (Global Anomaly Threshold to Unsupervised Detection), which can be used as an add-on component for any unsupervised anomaly detection approach to mitigate the sensitivity of such parameters whereby the performance of an unsupervised anomaly detection is improved. Experimental results show that GATUD has a significant improvement when compared to two well-known unsupervised anomaly detection methods.

6.1 INTRODUCTION

Two assumptions need to be made for unsupervised anomaly detection approaches: (i) the number of normal instances in data set vastly outperforms the abnormal instances and (ii) the abnormal instances must be statistically different from normal ones. Therefore, the performance of detection models relies mainly on assumptions to distinguish between normal and abnormal behavior. The reporting of anomalies in the unsupervised mode can be done either by scoring-based or binary-based methods.

In the scoring-based anomaly detection methods (Guttormsson et al. 1999; Eskin et al. 2002), all observations in a data set are given an anomaly score, where actual anomalies are assumed to have the highest scores. The key problem is how to find the best cut-off threshold that minimizes the false positive rate while maximizing the detection rate. On the one hand, binary-based methods (Portnoy et al., 2001; Mahoney and Chan 2003a) group similar observations together into a number of clusters. Abnormal observations are identified by making use of the fact that abnormal observations will be considered as outliers, and therefore will not be assigned to any

SCADA Security: Machine Learning Concepts for Intrusion Detection and Prevention, First Edition. Abdulmohsen Almalawi, Zahir Tari, Adil Fahad and Xun Yi. © 2021 John Wiley & Sons, Inc. Published 2021 by John Wiley & Sons, Inc.

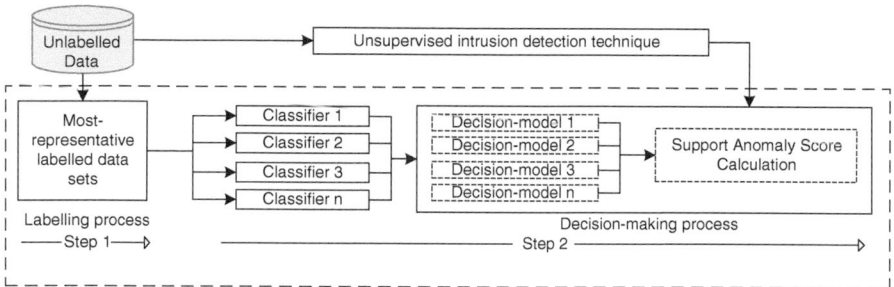

Figure 6.1 Overview of GATUD.

cluster, or they will be grouped into small clusters that have some characteristics that are different from normal clusters. However, labeling an observation as an outlier or a cluster as anomalous is controlled through some parameter choices within each detection method. For instance, given that the top 50% of the observations have the highest anomaly scores, these are assumed as outliers. In this case, both detection and false positive rates will be higher. Similarly, labelling a low percentage of largest clusters as normal in clustering-based intrusion detection methods, will result in higher detection and false positive rates. Therefore, the effectiveness of unsupervised intrusion approaches is sensitive to parameter choices, especially when the boundaries between normal and abnormal behaviors are not clearly distinguishable.

This chapter describes GATUD (Global Anomaly Threshold to Unsupervised Detection) that can be used as an add-on component to allow unsupervised anomaly scoring-based methods to set the value of the cut-off threshold parameter at a satisfactory level to guarantee a high detection rate, while minimizing the resulting high false positive rate. In addition, GATUD can be used as a robust method for labeling clusters to improve the accuracy of clustering-based intrusion detection systems. Figure 6.1 shows that GATUD involves two steps: (i) establishing two small-size most-representative data sets, where each data set represents one class problem (normal or abnormal) with high confidence and (ii) using the established data sets to build an ensemble-based decision-making model using a set of supervised classifiers.

This chapter is organized as follows. Section 6.2 presents an overview of related work and all details about the GATUD methods are described in Section 6.3. Section 6.4 presents the experimental set-up, followed by results and discussion in Section 6.5. Section 6.6 concludes the chapter.

6.2 RELATED WORK

As explained in earlier chapters, there are two categories of IDSs: signature-based and anomaly-based. The former detects only known attacks because it monitors the system against specific attack patterns. The latter attempts to build models from the normal behaviour of the systems, and any deviation from this behavior is assumed to be a malicious activity. Both approaches have advantages and disadvantages. The former achieves good accuracy, but fails to detect attacks that are new or the patterns of which are not learned. Although the latter is able to detect novel attacks, the overall detection accuracy of this approach is low.

This chapter focuses on the anomaly detection methods since they able to address the problem of the zero-day attacks. Rrushi et al. (2009a) applied statistics and probability theory to estimate the normality of the evolution of values of correlated process parameters. In the work of Valdes and Cheung (2009), the authors assumed that communication patterns among SCADA components are well-behaved, and combined the normal behavior of SCADA network traffic with artificial intrusion instances to learn the boundaries of the normal behavior using the neural network method. There are two types of anomaly detection methods: supervised and unsupervised modes. In the former mode, training data are labeled, while in the latter, data are not labeled. In contrast to conventional Information Technology (IT), the unsupervised mode has not been used much in SCADA systems. This is because SCADA security research is relatively recent compared with IT. In addition, the security requirements of such systems require a high detection accuracy, which this mode lacks.

Recently, machine learning methods have been successfully applied in traffic classification (Zhang et al., 2013a, 2013b) and unsupervised IDS (Portnoy et al., 2001; Mahoney and Chan, 2003a; Guttormsson et al., 1999; Eskin et al., 2002; MacQueen et al., 1967) for traditional IT networks. The ensemble-based clustering approaches enhance the performance of the unsupervised mode (Weng et al., 2007; Yamanishi and ichi Takeuchi, 2001) by combining classifiers to produce efficient and accurate IDSs (Kittler et al., 1998). Similar to the work carried out in this book, Yamanishi and ichi Takeuchi (2001) applied an anomaly-scoring method to unlabeled data, where the data that have the highest anomaly scores are labeled as outliers, while the rest are labeled as normal. They randomly selected a subset of normal data and combined it with outliers to create labeled data. Afterwards, a supervised method was trained with the labeled data to build an outlier filtering rule that differentiates outliers from normal data. The described GATUD (Global Anomaly Threshold to Unsupervised Detection) approach, however, differs in that the labeled data are learnt from data with no prior knowledge and a set of supervised classifiers are used to build a robust decision-maker because each classifier can capture different knowledge (Dieterich, 2000). Finally, GATUD is an add-on component (not an independent method) for unsupervised learning methods in order to benefit from the inherent characteristics of each algorithm.

6.3 GATUD APPROACH

The focus here is on improving the detection accuracy of unsupervised anomaly detection methods, as they are able to detect (unknown) zero-day attacks. However, these methods suffer from low accuracy. This section provides details about GATUD that are intended to address this problem. Below are the details of the various steps of this method.

6.3.1 Learning of Most-Representative Data Sets

In this step, two small-size, most-representative data sets are established from the unlabelled data, where the first and second data sets approximately represent the normal and abnormal behaviors, respectively. In order to choose the most-representative data sets, the following two steps are followed.

Step 1: Anomaly Scoring.   Since there is no prior knowledge about the normal and abnormal data, the k-nearest neighbor notion is adapted to assign an anomaly score to each observation. The k-nearest neighbour notion is chosen because it has produced significant results in anomaly scoring in the previous approaches (Guttormsson et al., 1999; Eskin et al., 2002) in cases where normal data in n-dimensional space form dense areas and the abnormal data are sparsely distributed. Unlike the previous approaches, we are concerned with the most relevant normal and abnormal data rather than with the detection of all anomalies. However, the k-nearest neighbor algorithm is computationally expensive, an issue that has been addressed in Chapter 4. Therefore, the kNNVWC approach is used to efficiently find k-nearest neighbors for each observation in a data set. Let D be an unlabeled data set of SCADA data with a multidimensional space, $m \times n$ matrix, where m and n represent the number of observations and attributes in D, respectively. Each dimension represents a distinct data point (e.g., temperature, motor speed, or humidity), while each observation x_i is represented by values of a set of attributes $A = \{a_1, a_2, a_3, \dots, a_n\}$. Let d be the Euclidian observation between two observations $x_1 = \{x_{1,1}, x_{1,2}, \dots, x_{1,n}\}$ and $x_2 = \{x_{2,1}, x_{2,2}, \dots, x_{2,n}\}$,

$$d(x_1, x_2) = \sqrt{\sum_{i=1}^{n} (x_{1,i} - x_{2,i})^2} \tag{6.1}$$

where n is the number of attributes. We compute the anomaly score for each observation x_i by taking the average of observations of its k-nearest neighbours. Let k be a positive parameter such that $2 \leq k \leq |D|$ and x_i be an observation in D. Then the anomaly score of x_i is computed as follows:

$$\rho(x_i) = \frac{1}{k} \sum_{i=0}^{k} D(x_i, knn_i(x_i)) \tag{6.2}$$

where $knn(x_i)$ is the k-nearest neighbors of observation x_i and the kNNVWC approach proposed in Chapter 3 is used here. Algorithm 6 summarises the calculation steps of anomaly scores for each observation x_i.

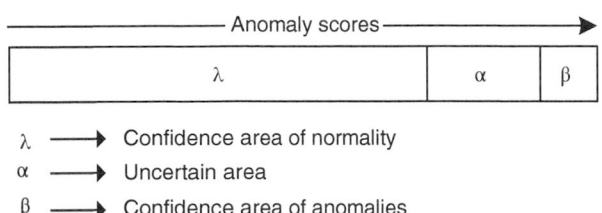

Figure 6.2   The categorization of unlabeled data after applying the anomaly-scoring method.

Algorithm 6   Anomaly scoring calculation

---

1 **Input:** D
 /* A matrix of unlabelled SCADA measurement data
    consisting of *m* observations and *n* attributes */
2 **Input:** k
 /* A positive integer that specifies the number of
    nearest neighbors   */
3 **Output:** AnomalyScoresList
 /* list of Anomaly Scores sorted by rank in
    descending order   */
4 *AnomalyScoresList* ⟵ Ø;
5 **foreach** *instance in D* **do**
6 │   *Score* ⟵ AnomalyScore (*instance*);
  │   /* Compute anomaly score as eq 6.2   */
7 │   put *Score* in *AnomalyScoresList*;
8 **return** [AnomalyScoresList];

Step 2: Selection of Candidate Sets.   From the list of anomaly scores, which are produced by Algorithm 6, two small-size, most-representative data sets are established where each data set represents normal or abnormal behaviors with high confidence. Based on the two previously mentioned assumptions of normal and abnormal behaviors in the unsupervised mode, the list of anomaly scores is grouped into three categories, as illustrated in Figure 6.2: *confidence area of anomalies, uncertain area*, and *confidence area of normality*. As shown in this figure, the extent of these areas is determined by the *confidence* thresholds β, α, and λ. For instance, the smaller the threshold β of the confidence area of anomalies, the greater is the confidence that the instances falling into this area are abnormal. This is true for the confidence area of the normality. Therefore, the thresholds (β and λ) should be kept at a distance from the uncertain area because this area requires a best cut-off threshold in order to judge whether the observation is either normal or abnormal, especially when some actual anomalous observations have anomaly scores that are close to some normal ones. Therefore, the most-representative data sets for normal and abnormal behaviors are established from the following two categories: *confidence area of* *normality* and *confidence area of anomalies*, respectively. The most-relevant anomalies, *AbnormalData*, are defined by selecting observations whose indices correspond to the top n of *AnomalyScoresList*, where $n = \beta \times |AnomalyScoresList|$, while the most relevant normal observations, *MostNormal*, are defined by selecting observations whose indices correspond to the bottom n of *AnomalyScoresList*, where $n = \lambda \times |AnomalyScoresList|$.

Again, if the two assumptions (Portnoy et al., 2001) about the unlabeled data are met, the thresholds β and λ are not difficult to determine. According to these assumptions, the anomalies are assumed to constitute a small portion of the data, where this percentage is assumed to not exceed 5%. Since we are not interested in finding all anomalies more than finding the fraction of anomalies with high confidence, and also we are not supposed to approach the uncertain area, the value of β will be set to a value that is smaller than 5%. As opposed to anomalies, the normal data is assumed to constitute a large portion of the data; therefore, setting λ to a small value will result in a small-size data set of most-relevant normal observations. However, this data set might not approximately represent the large portion of the normal data. To address this problem, λ is set here to a large value, providing this value overlaps with the uncertain area by, say, 80%. This will result in a large data set that is most-relevant normal. However, the computation time in the ensemble-based decision-making model will be substantially higher.

To resolve the previous problem, a small-size set of representative observations is extracted from the most relevant normal data set. During this step, similar observations are grouped together in terms of Euclidean distance, and take their mean as a representative observation for each group. The k-means clustering method (MacQueen et al., 1967) is a candidate algorithm for this process because of its simplicity, low computation time, and fast convergence. Moreover, the main disadvantages of k-means of determining the appropriate number of clusters and forcing an outlier observation to be assigned to the closest cluster even if it is dissimilar to its members will not be problematic in this step. This is because we are not interested in finding specific clusters more than chopping the data into a number of groups, and also the clustering data (most-relevant normal data set) is assumed to be outlier-free. Therefore, the number of clusters k will be set to a small value, where $k \ll |D|$. Algorithm 7 summarizes the steps involved in learning a small-size representative data set of the most relevant normal observations.

The two small-size data sets (the learned normal and anomalous data sets) are combined in order to form a labeled compressed representation of the unlabeled data. The concept of this compressed data set is slightly similar to the concept of the set of support vectors built by a Support Vector Machine (SVM) (Schiilköpf et al., 1995).

6.3.2 Decision-Making Model

This section introduces the *Ensemble-based Decision-Making Model* (EDMM) to compute the support anomaly score for each testing observation. As shown in Figure 6.1, EDMM is composed of a set of supervised classifiers whose individual

---

Algorithm 7   Learning the small-size representative data set of the most-relevant normal observations

---

1 **Input:** k
  /* A positive integer that specifies the number of
    clusters   */
2 **Input:** NormalData
  /* The instances that fall into the confidence area
    of normality   */
3 **Output:** RepNormalData
  /* small-size data set that contains representative
    observations for most-relevant normal
    observations   */
4 Initialise the cluster centroids $C = \{c_1, c_2, \ldots, c_k\}$;
5 *AssignmentList* $\longleftarrow \emptyset$;
  /* list of tuples < observation, Cluster ID >   */
6 **while** *termination criterion is not met* **do**
7   **foreach** *observation in NormalData* **do**
8   *ClusterID* $\longleftarrow$ ClosestCluster (*observation, C*) ;
        /* Find the closest cluster to this observation
                                                    */
9   put < *observation, ClusterID* > in *AssignmentList*;
10     $C \longleftarrow$ UpdateCentroids (*AssignmentList*) ;
11 *RepNormalData* $\longleftarrow C$;
  /* the centroids of clusters are used as
    representative observations for most-relevant
    normal observations   */
12 **return** [*RepNormalData*]];

---

decisions are combined to form an ensemble decision. This is because the combining of classifiers promised to be effective (Kittler et al., 1998). Each classifier c_i is trained with the labeled data set to build a decision model d_i. The number and type of involved supervised classifiers in GATUD have been left open because the choice of a specific algorithm is a critical step.

Let $C = \{c_i | 1 \leq i \geq n\}$ be a set of candidate supervised classifiers that build a set of decision models $D = \{d_i | 1 \leq i \geq n\}$. Each decision model d_i assigns a binary-decision value (either "1" or "0") to a testing observation x_i, $d_i(x_i) : v_i$. When the binary value v_i is "1", the observation x_i is judged as anomalous; otherwise is judged as normal. Then the calculation of the support anomaly score is defined as follows:

$$support(x_i) = \frac{\sum\limits_{j=1}^{n} d_j(x_i)}{n} \tag{6.3}$$

TABLE 6.1   Prediction Results for Decision Models on a Testing Observation x_i

Decision model	d_1	d_2	d_3	d_4	d_5	Sum
Is observation x_i anomalous?	1	1	0	1	1	4

where n is the number of decision models involved in the calculation of the support anomaly score. The observation type (class), whether abnormal or normal, is defined by the following equation:

$$Class(x_i) = \begin{cases} support(x_i) \geq \rho & \text{Abnormal} = 1 \\ Otherwise & \text{Normal} = 0 \end{cases} \tag{6.4}$$

where ρ is the percentage of the accepted vote of the decision models to judge a testing observation x_i as anomalous. For instance, when the threshold ρ is set to 1, the testing observation will not be considered as anomalous unless all involved decision models agree that the observation is an anomaly.

Illustrative Example.   A simple example is given below to illustrate the EDMM's process. Given five supervised classifiers are selected, $C = \{c_1, c_2, c_3, c_4, c_5\}$, and trained with labeled data sets that have been learned at Section 8, to build the decision models, $D = \{d_1, d_2, d_3, d_4, d_5\}$, and given a testing observation x_i whose status predicted by these in question models as shown in Table 6.1, then the support anomaly score is computed as follows:

$$support(x_i) = \frac{1 + 1 + 0 + 1 + 1}{5} = \frac{4}{5} = 0.80$$

Given the threshold ρ set to 0.6, where the observation x_i is considered as an anomaly, then at least three decision models have to assign it as anomalous. Therefore, from this example, the observation x_i will be considered anomalous.

6.4   EXPERIMENTAL SETUP

To provide quantitative results for GATUD, the seven labeled data sets provided in Chapter 5 are used to improve the accuracy and efficiency of the GATUD method.

6.4.1   Choice of Parameters

As discussed in Section 6.3.1, four parameters are needed to learn the most representative labeled data sets.

- The k-nearest neighbors parameter is the influencing factor for the anomaly scoring method. However, this value is insensitive and can be heuristically determined based on the assumption that anomalies constitute a tiny portion of the data. Therefore, the value of k-nearest is set to be 1% of the representative data set, because this value is assumed to discriminate between abnormal observations and normal ones in terms of the density-based distance.

- There are three parameters used for learning the most-representative data sets: (i) the extent of confidence area of normality λ; (ii) the extent of confidence area of abnormity β. The parameters λ and β are set to 70% and 1%, respectively. Even though the assumption was that the normal and abnormal data constitute larger percentages than the ones we have chosen, some distance from the uncertain area needs to be maintained; and (iii) the number of clusters k required for k-means in the candidate step. The purpose of this parameter is to reduce the number of representative normal observations, not to discover specific clusters. Experimentally, the value of k for several values such as 0.01%, 0.02%, 0.03%, and 0.04% demonstrated similar results, while the larger the value of k, the more is the computation time in the anomaly decision-making model of GATUD. Therefore, the value of this parameter is set to be 0.01% of the representative data set.

6.4.2 The Candidate Classifiers

As previously discussed, the type and the number of the supervised classifiers that are involved in EDMM in GATUD are left for the implementer. A thorough investigation has been conducted of a number of classifiers in this chapter, and finally the most five efficient classifiers have been choosen. Two are decision-tree based, best-first decision tree (BFTree) (Shi, 2007) and J48 (Quinlan, 1993); another two are rule-based, Non-Nested generalized exemplars (NNge) (Martin, 1995) and Projective Adaptive Resonance Theory (PART) (Frank and Witten, 1998); and the fifth is a probabilistic based, Naive Bayes (John and Langley, 1995). When using classifiers, we kept the default parameters of WEKA data mining software (Hall et al., 2009).

6.5 RESULTS AND DISCUSSION

Clearly, the GATUD method is intended to improve the accuracy of unsupervised anomaly detection methods in general and the SDAD method in particular. The evaluation provided here will demonstrate how GATUD can address limitations, where a global anomaly threshold is required to work with all data sets that vary in distribution, the number of inconsistent observations, and the application domain, when the *scoring-based* method is adapted. Furthermore, as mentioned earlier, GATUD can be used as an add-on component to help improve the accuracy of the unsupervised anomaly detection method. Its performance is demonstrated here when this is used with the k-means algorithm (MacQueen et al., 1967), which is considered to be one of the most useful and promising methods that can be adapted to build an unsupervised clustering-based anomaly detection method (Jianliang et al., 2009; Münz et al., 2007). The accuracy metrics that have been discussed and used in Chapter 5 are also used in this evaluation. Moreover, 10-fold cross-validation is applied to each data set for each experimental parameter.

6.5.1   Integrating GATUD into SDAD

The separation of the most relevant inconsistent observations from consistent ones in order to extract proximity detection rules for a given system is the initial part of the proposed data-driven clustering method (SDAD), described in Chapter 5. However, a cut-off threshold parameter η is required to be given, and in fact this parameter plays a major role in separating the most relevant inconsistent observations. The demonstrated results were significant; however, various cut-off thresholds η for a number of data sets have been used, where some data sets work with a small value of η, while others work with a large value. Therefore, we evaluate how the integration of GATUD into SDAD can help to find a global and efficient anomaly threshold η that can work with all data sets regardless of their variant characteristics, such as distribution, the number of inconsistent observations and the application domain, and meanwhile produces significant results.

It is well-known that the larger the value of the cut-off threshold η, the higher will be the detection rate and the higher will be the false positive rate as well. This, however, will result in poor performance. The determination of an appropriate cut-off threshold η that maximizes and minimizes the detection rate and the false positive rate, respectively, is the challenging problem. GATUD addresses this problem by allowing the anomaly scoring method to choose a large value of cut-off thresholds η in order to ensure that the detection rate is higher, while minimizing the false positive rate without degrading the detection rate.

In what follows will show the separation accuracy results with/without the integration of GATUD into SDAD. Then, we demonstrate how this integration has a significant impact on the accuracy of the generation step of proximity detection rules, which is the second phase following the separation process.

Results of the Separation Process With/Without GATUD,   In particular, the results that meet the following criteria are considered as significant:

$$\begin{cases} Detection\ rate \geq 90\% \\ False\ positive\ rate \leq 1.5\% \end{cases} \tag{6.5}$$

Tables 6.2 to 6.8 show the separation accuracy results with/without the integration of GATUD. Clearly, as shown in the result tables, the larger the value of the cut-off threshold η, the higher the detection rate of inconsistent observations. This is because the observations are sorted by their anomaly scores in ascending order and the consideration of the top large portion of the sorted list increases the chance to obtain the actual inconsistent observations. However, this will result in a large number of consistent observations existing in this portion, which definitely increases the false positive

TABLE 6.2 **The Separation Accuracy of Inconsistent Observations With/Without GATUD on DUWWTP**

η	Without GATUD				With GATUD			
	DR	FPR	P	F-M	DR	FPR	P	F-M
0.50%	14.29%	0.00%	100.00%	25.00%	14.29%	0.00%	100.00%	25.00%
1.00%	35.71%	0.00%	100.00%	52.63%	35.71%	0.00%	100.00%	52.63%
1.50%	50.00%	0.00%	100.00%	66.67%	50.00%	0.00%	100.00%	66.67%
2.00%	70.71%	0.02%	99.00%	82.50%	64.29%	0.02%	98.90%	77.92%
2.50%	79.29%	0.19%	92.50%	85.38%	71.43%	0.13%	94.34%	81.30%
3.00%	87.14%	0.39%	87.14%	87.14%	78.57%	0.13%	94.83%	85.94%
3.50%	**92.86%**	**0.87%**	76.47%	83.87%	**92.86%**	**0.13%**	95.59%	94.20%
4.00%	**92.86%**	**1.30%**	68.42%	78.79%	**92.86%**	**0.28%**	90.91%	91.87%
4.50%	94.29%	1.69%	62.86%	75.43%	**92.86%**	**0.43%**	86.67%	89.66%
5.00%	100.00%	2.17%	58.33%	73.68%	**92.86%**	**0.54%**	83.87%	88.14%

η: Top percentage of observations, which are sorted by inconsistency scores in ascending order, are assumed as inconsistent.
DR: Detection Rate.
FPR: False Positive Rate.
P: Precision.
$F-M$: F-Measure.

TABLE 6.3 **The Separation Accuracy of Inconsistent Observations With/Without GATUD on SimData1**

η	Without GATUD				With GATUD			
	DR	FPR	P	F-M	DR	FPR	P	F-M
0.50%	40.78%	0.06%	88.51%	55.84%	40.78%	0.06%	88.51%	55.84%
1.00%	80.78%	0.13%	86.74%	83.65%	80.78%	0.13%	86.74%	83.65%
1.50%	**98.04%**	**0.45%**	70.42%	81.97%	**98.04%**	**0.25%**	80.91%	88.65%
2.00%	**98.04%**	**0.95%**	52.91%	68.73%	**98.04%**	**0.41%**	72.05%	83.06%
2.50%	**98.04%**	**1.46%**	42.19%	59.00%	**98.04%**	**0.44%**	70.87%	82.27%
3.00%	98.04%	1.97%	35.21%	51.81%	**98.04%**	**0.46%**	69.78%	81.53%
3.50%	98.04%	2.47%	30.21%	46.19%	**98.04%**	**0.48%**	68.82%	80.87%
4.00%	98.04%	2.97%	26.46%	41.67%	**98.04%**	**0.50%**	68.17%	80.42%
4.50%	98.04%	3.48%	23.47%	37.88%	**98.04%**	**0.52%**	67.48%	79.94%
5.00%	98.04%	3.99%	21.14%	34.78%	**98.04%**	**0.52%**	67.29%	79.81%

rate. On the another hand, it can be seen that the use of GATUD can benefit from the larger value of the cut-off threshold η in order to maximize the detection rate of inconsistent observations and the obtained list of the assumed inconsistent observations is passed through the decision-making model to rejudge whether each observation is inconsistent or consistent. This, as can be seen in the result tables, nearly sustains the detection rate and meanwhile minimizes the false positive rate.

TABLE 6.4 **The Separation Accuracy of Inconsistent Observations With/Without GATUD on SimData2**

η	Without GATUD				With GATUD			
	DR	FPR	P	F-M	DR	FPR	P	F-M
0.50%	46.50%	0.01%	98.94%	63.27%	46.50%	0.01%	98.94%	63.27%
1.00%	81.50%	0.14%	85.79%	83.59%	81.50%	0.14%	85.79%	83.59%
1.50%	**100.00%**	**0.45%**	70.42%	82.64%	**100.00%**	**0.16%**	86.88%	92.98%
2.00%	**100.00%**	**0.95%**	52.91%	69.20%	**100.00%**	**0.19%**	84.82%	91.79%
2.50%	**100.00%**	**1.46%**	42.19%	59.35%	**100.00%**	**0.23%**	82.10%	90.17%
3.00%	100.00%	1.97%	35.21%	52.08%	**100.00%**	**0.26%**	80.71%	89.33%
3.50%	100.00%	2.47%	30.21%	46.40%	**100.00%**	**0.27%**	80.06%	88.93%
4.00%	100.00%	2.97%	26.46%	41.84%	**100.00%**	**0.28%**	79.37%	88.50%
4.50%	100.00%	3.48%	23.47%	38.02%	**100.00%**	**0.29%**	78.43%	87.91%
5.00%	100.00%	3.98%	21.14%	34.90%	**100.00%**	**0.31%**	77.40%	87.26%

TABLE 6.5 **The Separation Accuracy of Inconsistent Observations With/Without GATUD on SIRD**

η	Without GATUD				With GATUD			
	DR	FPR	P	F-M	DR	FPR	P	F-M
0.50%	17.09%	0.00%	100.00%	29.20%	17.09%	0.00%	100.00%	29.20%
1.00%	34.19%	0.00%	100.00%	50.96%	34.19%	0.00%	100.00%	50.96%
1.50%	51.28%	0.00%	100.00%	67.80%	51.28%	0.00%	100.00%	67.80%
2.00%	68.38%	0.00%	100.00%	81.22%	68.38%	0.00%	100.00%	81.22%
2.50%	85.47%	0.00%	100.00%	92.17%	85.47%	0.00%	100.00%	92.17%
3.00%	**100.00%**	**0.08%**	97.50%	98.73%	**100.00%**	**0.08%**	97.50%	98.73%
3.50%	**100.00%**	**0.59%**	83.57%	91.05%	**100.00%**	**0.52%**	85.21%	92.02%
4.00%	**100.00%**	**1.09%**	73.58%	84.78%	**100.00%**	**0.85%**	78.10%	87.71%
4.50%	100.00%	1.60%	65.36%	79.05%	**100.00%**	**0.97%**	75.73%	86.19%
5.00%	100.00%	2.12%	58.79%	74.05%	**100.00%**	**0.99%**	75.24%	85.87%

Table 6.2 shows that the significant accuracy results of separation of inconsistent observations without the integration of GATUD are achieved when the cut-off threshold η is set with 3.5% and 4.0%. Moreover, it can be observed that the setting of η with larger values such as 4.5% and 5.0% has not demonstrated any significant results even though the detection rates of the inconsistent observations were high. This is because, as demonstrated, the false positive rates were relatively high. On the other hand, when GATUD was integrated, the high false positive rates were significantly reduced and meanwhile the detection rates were sustained at significantly acceptable levels. Similarly, the remaining results for each data set (as shown in Tables 6.3 to 6.8) demonstrated that the integration of GATUD significantly reduced the high false positive rates when larger values of η were used, and meanwhile maintained the

TABLE 6.6   The Separation Accuracy of Inconsistent Observations With/Without GATUD on SORD

η	Without GATUD				With GATUD			
	DR	FPR	P	F-M	DR	FPR	P	F-M
0.50%	71.88%	0.00%	100.00%	83.64%	71.88%	0.00%	100.00%	83.64%
1.00%	**90.94%**	**0.35%**	64.67%	75.58%	**90.63%**	**0.35%**	64.59%	75.42%
1.50%	**93.75%**	**0.84%**	44.12%	60.00%	**90.63%**	**0.59%**	52.06%	66.13%
2.00%	93.75%	1.77%	27.27%	42.25%	**90.63%**	**0.70%**	47.93%	62.70%
2.50%	93.75%	2.00%	25.00%	39.47%	**90.63%**	**0.72%**	47.15%	62.03%
3.00%	96.88%	2.33%	22.79%	36.90%	**90.63%**	**0.72%**	47.08%	61.97%
3.50%	96.88%	2.84%	19.50%	32.46%	**90.63%**	**0.72%**	47.08%	61.97%
4.00%	96.88%	3.35%	17.03%	28.97%	**90.63%**	**0.72%**	47.08%	61.97%
4.50%	96.88%	3.84%	15.20%	26.27%	**90.63%**	**0.72%**	47.08%	61.97%
5.00%	96.88%	4.35%	13.66%	23.94%	**90.63%**	**0.72%**	47.08%	61.97%

TABLE 6.7   The Separation Accuracy of Inconsistent Observations With/Without GATUD on MORD

η	Without GATUD				With GATUD			
	DR	FPR	P	F-M	DR	FPR	P	F-M
0.50%	36.21%	0.00%	100.00%	53.16%	36.21%	0.00%	100.00%	53.16%
1.00%	72.41%	0.00%	100.00%	84.00%	72.41%	0.00%	100.00%	84.00%
1.50%	86.03%	0.31%	79.21%	82.48%	75.86%	0.00%	100.00%	86.27%
2.00%	86.21%	0.84%	58.82%	69.93%	75.86%	0.00%	100.00%	86.27%
2.50%	86.55%	1.34%	47.36%	61.22%	75.86%	0.00%	100.00%	86.27%
3.00%	87.41%	1.83%	39.92%	54.81%	75.86%	0.00%	100.00%	86.27%
3.50%	87.93%	2.33%	34.46%	49.51%	75.86%	0.00%	100.00%	86.27%
4.00%	87.93%	2.83%	30.18%	44.93%	75.86%	0.00%	100.00%	86.27%
4.50%	88.45%	3.33%	27.00%	41.37%	75.86%	0.00%	100.00%	86.27%
5.00%	89.31%	3.82%	24.55%	38.51%	75.86%	0.00%	100.00%	86.27%

detection rates at a satisfactory and significant level. However, the results for the data set *MORD* in Table 6.7 were not significant whether GATUD was integrated or not. We would have expected the integration of GATUD to produce significant results if the cut-off threshold η was set to a value that is greater than 0.05%. However, this value is assumed as the maximum percentage of inconsistent observations in an unlabeled data set. Therefore, this data set is considered to be an exceptional case.

Results of Proximity Detection Rules With/Without GATUD

As mentioned previously, the generation process of proximity detection rules comes after and relies on the separation process. Therefore, the robustness of these proximity

TABLE 6.8   The Separation Accuracy of Inconsistent Observations With/Without GATUD on MIRD

η	Without GATUD				With GATUD			
	DR	FPR	P	F-M	DR	FPR	P	F-M
0.50%	21.00%	0.00%	100.00%	34.71%	21.00%	0.00%	100.00%	34.71%
1.00%	42.00%	0.00%	100.00%	59.15%	42.00%	0.00%	100.00%	59.15%
1.50%	63.00%	0.00%	100.00%	77.30%	63.00%	0.00%	100.00%	77.30%
2.00%	85.00%	0.00%	100.00%	91.89%	85.00%	0.00%	100.00%	91.89%
2.50%	**100.00%**	**0.15%**	94.34%	97.09%	**97.00%**	**0.00%**	100.00%	98.48%
3.00%	**100.00%**	**0.65%**	78.74%	88.11%	**97.00%**	**0.00%**	100.00%	98.48%
3.50%	**100.00%**	**1.16%**	67.57%	80.65%	**97.00%**	**0.00%**	100.00%	98.48%
4.00%	100.00%	1.67%	59.17%	74.35%	**97.00%**	**0.00%**	100.00%	98.48%
4.50%	100.00%	2.18%	52.63%	68.97%	**97.00%**	**0.00%**	100.00%	98.48%
5.00%	100.00%	2.71%	47.17%	64.10%	**97.00%**	**0.00%**	100.00%	98.48%

detection rules is influenced by the accuracy of the separation process and, as earlier shown, the integration of GATUD demonstrated significant results in the separation process, even with large cut-off thresholds η. Therefore, the detection accuracy results of the proximity detection rules, which are extracted from the inconsistent and consistent observations that were separated using such these large cut-off thresholds η, are expected to be significant. Tables 6.9 to 6.15 show the detection accuracy results.

Each table represents the results of the detection accuracy results for each individual data set, and are also divided into two parts: the first part shows the results of the proximity-detection rules that were extracted by separating inconsistent from consistent observations where GATUD was not integrated in the separation process. The second part shows the results obtained after the integration of GATUD. The result tables show that the integration of GATUD into the separation process helps to generate robust proximity-detection rules, even with large cut-off thresholds η.

Overall, Table 6.16 highlights the acceptable thresholds η through which the extracted proximity-detection rules demonstrated significant detection accuracy results, where GATUD was integrated into the separation process of inconsistent and consistent observations. From this table, the determination of the near-optimal value of a cut-off threshold η will not be problematic because the value of η can be set to 0.05, which is assumed as the maximum percentage of inconsistent observations in an unlabeled data set. The resultant high positive rate that might result from this large value can significantly be reduced by the integration of GATUD.

6.5.2   Integrating GATUD into the Clustering-based Method

This section shows how GATUD can be integrated not only with the scoring-based intrusion detection method but also with the clustering-based method. The k-means algorithm, which is considered as one of the most useful and promising methods

TABLE 6.9 The Detection Accuracy of the Proximity-Detection Rules That Have Been Extracted With/Without the Integration of GATUD in the Separation Process on DUWWTP

η	w	Without GATUD						With GATUD					
		NC	AC	DR	FPR	P	F-M	NC	AC	DR	FPR	P	F-M
0.50%	0.7506	62	2	14.29%	0.00%	100.00%	25.00%	62	2	14.29%	0.00%	100.00%	25.00%
1.00%		59	5	35.71%	0.00%	100.00%	52.63%	59	5	35.71%	0.00%	100.00%	52.63%
1.50%		57	7	50.00%	0.00%	100.00%	66.67%	57	7	50.00%	0.00%	100.00%	66.67%
2.00%		54	10	70.71%	0.00%	100.00%	82.85%	55	9	64.29%	0.00%	100.00%	78.26%
2.50%		52	12	79.29%	0.00%	100.00%	88.45%	54	10	71.43%	0.00%	100.00%	83.33%
3.00%		51	14	87.14%	5.85%	80.26%	83.56%	53	11	78.57%	0.00%	100.00%	88.00%
3.50%		48	16	92.86%	0.78%	97.01%	**94.89%**	52	12	84.29%	0.00%	100.00%	**91.47%**
4.00%		46	18	92.86%	0.97%	96.30%	**94.55%**	52	12	84.29%	0.00%	100.00%	**91.47%**
4.50%		44	20	94.29%	9.75%	72.53%	81.99%	51	13	85.71%	0.19%	99.17%	**91.95%**
5.00%		41	23	100.00%	11.70%	70.00%	82.35%	49	15	89.29%	0.58%	97.66%	**93.28%**

η: Top percentage of observations, which are sorted by inconsistency scores in ascending order, are assumed as inconsistent.
w: The cluster width parameter.
NC: The number of the produced normal clusters.
AC: The number of the produced abnormal clusters.
DR: Detection Rate.
FPR: False Positive Rate.
P: Precision.
F-M: F-Measure.

TABLE 6.10 The Detection Accuracy of the Proximity-Detection Rules That Have Been Extracted With/Without the Integration of GATUD in the Separation Process on SimData1

η	w	Without GATUD						With GATUD					
		NC	AC	DR	FPR	P	F-M	NC	AC	DR	FPR	P	F-M
0.50%	0.1456	316	39	40.78%	0.06%	98.58%	57.70%	316	39	40.78%	0.06%	98.58%	57.70%
1.00%		286	68	80.78%	0.13%	98.33%	88.70%	286	68	80.78%	0.13%	98.33%	88.70%
1.50%		265	92	98.04%	0.46%	95.42%	96.71%	272	81	98.04%	0.28%	97.18%	97.61%
2.00%		252	104	98.04%	1.11%	89.69%	93.68%	269	82	98.04%	0.42%	95.79%	96.90%
2.50%		242	116	98.04%	1.73%	84.75%	90.91%	269	82	98.04%	0.44%	95.60%	96.81%
3.00%		238	131	98.04%	2.49%	79.43%	87.76%	268	85	98.04%	0.48%	95.24%	96.62%
3.50%		234	138	98.04%	3.08%	75.76%	85.47%	267	85	98.04%	0.56%	94.52%	96.25%
4.00%		225	144	98.04%	3.64%	72.52%	83.37%	267	85	98.04%	0.58%	94.34%	96.15%
4.50%		219	151	98.04%	4.24%	69.40%	81.27%	266	85	98.04%	0.59%	94.25%	96.11%
5.00%		211	157	98.04%	4.75%	66.93%	79.55%	267	85	98.04%	0.59%	94.25%	96.11%

TABLE 6.11 The Detection Accuracy of the Proximity-Detection Rules That Have Been Extracted With/Without the Integration of GATUD in the Separation Process on SimData2

η	w	Without GATUD						With GATUD					
		NC	AC	DR	FPR	P	F-M	NC	AC	DR	FPR	P	F-M
0.50%	0.1476	307	42	46.50%	0.01%	99.79%	63.44%	307	42	46.50%	0.01%	99.79%	63.44%
1.00%		279	71	70.00%	0.14%	97.90%	81.63%	279	71	70.00%	0.14%	97.90%	81.63%
1.50%		262	90	100.00%	0.51%	94.97%	97.42%	268	78	100.00%	0.17%	98.23%	99.11%
2.00%		255	102	100.00%	1.18%	89.05%	94.21%	268	78	100.00%	0.20%	97.94%	98.96%
2.50%		243	112	100.00%	1.72%	84.82%	91.79%	268	78	100.00%	0.26%	97.37%	98.67%
3.00%		239	120	100.00%	2.42%	79.87%	88.81%	268	78	100.00%	0.26%	97.37%	98.67%
3.50%		231	130	100.00%	3.05%	75.93%	86.32%	268	78	100.00%	0.26%	97.37%	98.67%
4.00%		228	139	100.00%	3.63%	72.57%	84.10%	268	78	100.00%	0.28%	97.18%	98.57%
4.50%		222	146	100.00%	4.19%	69.64%	82.10%	268	78	100.00%	0.31%	96.90%	98.43%
5.00%		216	147	100.00%	4.71%	67.11%	80.32%	268	78	100.00%	0.37%	96.34%	98.14%

TABLE 6.12 The Detection Accuracy of the Proximity-Detection Rules That Have Been Extracted With/Without the Integration of GATUD in the Separation Process on SIRD

η	w	Without GATUD						With GATUD					
		NC	AC	DR	FPR	P	F-M	NC	AC	DR	FPR	P	F-M
0.50%	0.0124	63	17	17.09%	0.00%	100.00%	29.20%	63	17	17.09%	0.00%	100.00%	29.20%
1.00%		45	35	34.19%	0.00%	100.00%	50.96%	45	35	34.19%	0.00%	100.00%	50.96%
1.50%		36	44	51.97%	0.00%	100.00%	68.39%	36	44	51.97%	0.00%	100.00%	68.39%
2.00%		31	49	67.61%	0.00%	100.00%	80.67%	31	49	67.61%	0.00%	100.00%	80.67%
2.50%		24	57	72.65%	0.00%	100.00%	84.16%	24	57	70.09%	0.00%	100.00%	82.41%
3.00%		17	63	100.00%	0.42%	98.48%	**99.24%**	17	63	100.00%	0.42%	98.48%	**99.24%**
3.50%		15	64	100.00%	0.77%	97.26%	**98.61%**	16	63	100.00%	0.65%	97.66%	**98.82%**
4.00%		15	64	100.00%	1.58%	94.51%	**97.18%**	16	63	100.00%	0.84%	97.01%	**98.48%**
4.50%		15	64	100.00%	6.98%	79.59%	88.64%	16	63	100.00%	0.93%	96.69%	**98.32%**
5.00%		15	65	100.00%	8.14%	76.97%	86.99%	16	64	100.00%	0.93%	96.69%	**98.32%**

TABLE 6.13 The Detection Accuracy of the Proximity-Detection Rules That Have Been Extracted With/Without the Integration of GATUD in the Separation Process on SORD

η	w	Without GATUD						With GATUD					
		NC	AC	DR	FPR	P	F-M	NC	AC	DR	FPR	P	F-M
0.50%	0.0152	73	20	71.88%	0.00%	100.00%	83.64%	73	20	71.88%	0.00%	100.00%	83.64%
1.00%		67	27	90.94%	0.28%	95.41%	**93.12%**	67	27	90.63%	0.28%	95.39%	**92.95%**
1.50%		64	32	93.75%	0.90%	86.96%	**90.23%**	67	28	90.63%	0.60%	90.63%	**90.63%**
2.00%		62	33	93.75%	1.32%	81.97%	87.46%	67	28	90.63%	0.72%	88.96%	**89.78%**
2.50%		61	35	93.75%	2.02%	74.81%	83.22%	67	28	90.63%	0.74%	88.69%	**89.64%**
3.00%		57	37	96.88%	2.38%	72.26%	82.78%	67	28	90.63%	0.74%	88.69%	**89.64%**
3.50%		55	41	96.88%	3.03%	67.10%	79.28%	67	28	90.63%	0.74%	88.69%	**89.64%**
4.00%		55	42	96.88%	3.83%	61.75%	75.43%	67	28	90.63%	0.74%	88.69%	**89.64%**
4.50%		54	41	96.88%	4.19%	59.62%	73.81%	67	28	90.63%	0.74%	88.69%	**89.64%**
5.00%		54	42	96.88%	4.59%	57.41%	72.09%	67	28	90.63%	0.74%	88.69%	**89.64%**

TABLE 6.14 The Detection Accuracy of the Proximity-Detection Rules That Have Been Extracted With/Without the Integration of GATUD in the Separation Process on MORD

η	w	Without GATUD						With GATUD					
		NC	AC	DR	FPR	P	F-M	NC	AC	DR	FPR	P	F-M
0.50%	0.0093	74	14	36.21%	0.00%	100.00%	53.16%	74	14	36.21%	0.00%	100.00%	53.16%
1.00%		55	34	72.59%	0.00%	100.00%	84.12%	55	34	72.59%	0.00%	100.00%	84.12%
1.50%		53	38	77.59%	0.56%	94.54%	85.23%	55	34	77.41%	0.00%	100.00%	87.27%
2.00%		53	39	86.21%	1.10%	90.74%	88.42%	55	34	77.41%	0.00%	100.00%	87.27%
2.50%		52	39	86.55%	1.68%	86.55%	86.55%	55	34	77.41%	0.00%	100.00%	87.27%
3.00%		50	40	87.41%	2.20%	83.25%	85.28%	55	34	77.41%	0.00%	100.00%	87.27%
3.50%		49	41	87.93%	2.72%	80.19%	83.88%	55	34	77.41%	0.00%	100.00%	87.27%
4.00%		48	42	87.93%	3.00%	78.58%	82.99%	55	34	77.41%	0.00%	100.00%	87.27%
4.50%		48	43	88.45%	3.39%	76.57%	82.08%	55	34	77.41%	0.00%	100.00%	87.27%
5.00%		46	44	89.31%	3.99%	73.68%	80.75%	55	34	77.41%	0.00%	100.00%	87.27%

TABLE 6.15 The Detection Accuracy of the Proximity-Detection Rules That Have Been Extracted With/Without the Integration of GATUD in the separation process on MIRD

η	w	Without GATUD						With GATUD					
		NC	AC	DR	FPR	P	F-M	NC	AC	DR	FPR	P	F-M
0.50%	0.0135	62	13	20.10%	0.00%	100.00%	33.47%	62	13	20.10%	0.00%	100.00%	33.47%
1.00%		44	31	42.00%	0.00%	100.00%	59.15%	44	31	42.00%	0.00%	100.00%	59.15%
1.50%		37	38	63.00%	0.00%	100.00%	77.30%	37	38	63.00%	0.00%	100.00%	77.30%
2.00%		26	49	70.00%	0.00%	100.00%	82.35%	26	49	70.00%	0.00%	100.00%	82.35%
2.50%		20	55	100.00%	0.20%	99.11%	**99.55%**	22	54	96.90%	0.00%	100.00%	**98.43%**
3.00%		18	57	100.00%	0.87%	96.15%	**98.04%**	22	54	96.90%	0.00%	100.00%	**98.43%**
3.50%		18	58	100.00%	1.29%	94.43%	**97.13%**	22	54	96.90%	0.00%	100.00%	**98.43%**
4.00%		18	59	100.00%	6.54%	76.92%	86.96%	22	54	96.90%	0.00%	100.00%	**98.43%**
4.50%		15	61	100.00%	7.63%	74.07%	85.11%	22	54	96.90%	0.00%	100.00%	**98.43%**
5.00%		16	65	100.00%	8.28%	72.46%	84.03%	22	54	96.90%	0.00%	100.00%	**98.43%**

TABLE 6.16   The Acceptable Thresholds η that Produce Significant accuracy results of the Separation of Inconsistent Observations on Each Data Set When GATUD Was Intergraded

η	Data sets							AgreeNO
	DUWWTP	SimData1	SimData2	MORD	MIRD	SORD	SIRD	
0.50%								0
1.00%					✓			1
1.50%		✓	✓		✓			3
2.00%		✓	✓		✓			3
2.50%		✓	✓		✓		✓	4
3.00%		✓	✓	✓	✓		✓	5
3.50%	✓	✓	✓	✓	✓		✓	6
4.00%	✓	✓	✓	✓	✓		✓	6
4.50%	✓	✓	✓	✓	✓		✓	6
5.00%	✓	✓	✓	✓	✓		✓	6

η: Top percentage of observations, that are sorted by anomaly scores in ascending order, are assumed as inconsistent # of Agreement: The number of data sets that agree on each separation threshold η, where the agreement is judged by the significant detection results

for building an unsupervised clustering-based intrusion detection model (Jianliang et al., 2009; Münz et al., 2007), is chosen to demonstrate the integration effectiveness of GATUD with an unsupervised clustering-based intrusion detection method. Therefore, it is interesting to demonstrate detection accuracy results with/without the integration of GATUD. For more details about how this algorithm can be adapted to build an unsupervised anomaly detection method, see Almalawi et al. (2015).

In the adaptation of k-means for building an unsupervised inconsistent detection model, anomalies are assumed to be grouped in clusters that contain percentage θ of the data. Let $C = C_1, C_2, \dots C_n$ be the set of clusters that have been created. Then the anomalous clusters are defined as follows:

$$\acute{C} = \{\acute{C}_1, \acute{C}_2, \dots, \acute{C}_b\} = \sum_{i=1}^{n} |C_i| \leq \theta \tag{6.6}$$

while the remaining clusters $C - \acute{C}$ are labeled as normal. In this evaluation, we assume that the real anomalous cluster is the cluster where the majority of its members are actual inconsistent observations. Assume that the number of inconsistent observations $\geq |\acute{C}_i|/2$. Therefore, the labeling accuracy of the assumed percentage θ of the data in anomalous clusters is measured by the Labeling Error Rate (LER) for clusters:

$$LER = \frac{Number\ of\ clusters\ incorrectly\ labeled}{Number\ of\ all\ identified\ clusters} \tag{6.7}$$

When integrating GATUD to label the clusters, the members of each individual cluster pass through the decision-making model to be labeled as either consistent or inconsistent. Then the cluster is labeled according to the label of the majority of its members. The labeling of clusters by GATUD is given as follows:

$$L(C_i) = \sum_{j=1}^{|C_i|} Class(x_j) \tag{6.8}$$

where $L(C_i)$ is the number of inconsistent observations, which are judged by the decision-making model, in the cluster C_i. Then the anomalous clusters are defined as follows:

$$\acute{C} = \{\acute{C}_1, \acute{C}_2, \ldots, \acute{C}_b\} = \sum_{i=1}^{n} L(C_i) \geq \varepsilon \times |C_i| \tag{6.9}$$

where ε is the percentage of inconsistent observations in a cluster C_i to be labeled as inconsistent. In this evaluation, it is set to 0.5.

The integration of GATUD is evaluated as an add-in component with k-means, where this component is only used to label the produced clusters as either consistent or inconsistent. The k-means requires two user-specified parameters k and θ to build the unsupervised anomaly detection model from unlabeled data, where k is the number of clusters and θ is the percentage of the data in a cluster to be malicious. However, the parameter θ is not required when GATUD is integrated. This evaluation shows the detection accuracy of k-means as an independent/dependent algorithm. In the independent use, k-means is used to cluster the training data set and labels each cluster using an assumption of the percentage of the data in a cluster to be assumed as malicious. While in the dependent use, GATUD is used as a labeling method for the produced clusters by k-means. The parameters k and θ are set to the same values that have been used in the paper by Almalawi et al. (2015). This is because the same data sets are used.

Tables 6.17 to 6.23 show the detection accuracy results of the detection model that was built by the clustering k-means algorithm, and, as shown, these results represent the independent and dependent use of k-means, where, in the latter, GATUD was integrated to label the clusters. This evaluation showed only the results of F-measure, as they are the interesting results to compare. Clearly, the detection accuracy results of k-means in detecting inconsistent observations are very poor for all data sets when GATUD was not integrated. On the other hand, significant results for some data sets are obtained when GATUD is integrated to label the produced clusters. It is obvious from the results that GATUD can be a promising method to improve the accuracy of an unsupervised anomaly detection approach, not only with the SDAD approach proposed by Almalawi et al. (2015), but also where it can be integrated with unsupervised clustering-based anomaly detection models.

6.6   CONCLUSION

This chapter described an innovative method, namely GATUD (Global Anomaly Threshold to Unsupervised Detection), which is used as an add-on component to

TABLE 6.17 The Detection Accuracy of *k*-Means Clustering Algorithm With/Without GATUD on DUWWTP

K	Without GATUD										With GATUD	
	$\theta = 0.01$		$\theta = 0.02$		$\theta = 0.03$		$\theta = 0.04$		$\theta = 0.05$			
	LER	F-M	LER	F-M	LER	F-M	LER	F-M	LER	F-M	LER	F-M
50	12.40%	69.65%	29.30%	59.98%	43.73%	50.50%	53.80%	44.22%	60.92%	39.86%	6.00%	79.37%
60	20.17%	62.36%	39.08%	53.36%	53.11%	44.37%	61.92%	39.03%	67.63%	35.62%	7.17%	74.14%
70	23.57%	62.95%	46.43%	48.73%	59.76%	40.56%	67.39%	36.05%	72.20%	33.17%	6.71%	79.36%
80	33.75%	57.08%	54.63%	44.19%	66.25%	37.23%	72.75%	33.28%	76.73%	30.91%	5.88%	79.42%
90	40.44%	58.62%	62.17%	42.73%	71.96%	35.87%	77.17%	32.26%	80.27%	30.10%	7.11%	80.28%
100	47.60%	51.52%	67.10%	38.56%	75.63%	33.01%	80.08%	30.13%	82.74%	28.39%	6.40%	72.13%

DR: Labeling error-rate.
F-M: F-Measure.
θ: The percentage of the data in a cluster assumed to be malicious.
K: The number of clusters.

TABLE 6.18 The Detection Accuracy of *k*-Means Clustering Algorithm With/Without GATUD on SimData1

| | Without GATUD | | | | | | | | With GATUD | |
| | $\theta = 0.01$ | | $\theta = 0.02$ | | $\theta = 0.03$ | | $\theta = 0.04$ | | $\theta = 0.05$ | | | |
K	LER	F-M	LER	F-M	LER	F-M	LER	F-M	LER	F-M	LER	F-M
50	10.40%	50.56%	33.40%	30.82%	50.00%	23.74%	61.10%	19.79%	68.48%	17.67%	0.60%	72.59%
60	18.50%	37.04%	42.67%	24.17%	58.83%	19.06%	68.46%	15.51%	74.37%	14.23%	0.00%	68.50%
70	28.86%	26.97%	55.29%	18.66%	68.81%	15.02%	76.00%	13.32%	80.51%	12.44%	0.29%	64.13%
80	40.88%	30.68%	64.69%	20.10%	75.58%	16.30%	81.19%	14.51%	84.68%	13.39%	0.13%	**98.11%**
90	41.67%	24.61%	67.56%	17.29%	77.93%	14.50%	83.14%	12.69%	86.20%	11.94%	0.00%	86.28%
100	53.90%	21.15%	75.05%	15.34%	82.90%	13.22%	86.88%	11.96%	89.28%	11.35%	0.40%	86.77%

TABLE 6.19 The Detection Accuracy of k-Means Clustering Algorithm With/Without GATUD on SimData2

K	Without GATUD $\theta = 0.01$		$\theta = 0.02$		$\theta = 0.03$		$\theta = 0.04$		$\theta = 0.05$		With GATUD	
	LER	F-M	LER	F-M	LER	F-M	LER	F-M	LER	F-M	LER	F-M
50	10.80%	74.79%	31.70%	48.40%	48.60%	36.25%	59.95%	29.58%	67.24%	25.49%	0.60%	99.71%
60	21.00%	52.54%	44.58%	33.94%	59.89%	26.06%	69.17%	21.77%	74.80%	19.17%	0.33%	99.61%
70	25.86%	43.87%	52.07%	28.49%	66.67%	22.10%	74.29%	18.80%	78.89%	16.80%	0.14%	99.80%
80	35.88%	33.18%	61.56%	22.04%	73.17%	17.76%	79.25%	15.55%	82.98%	14.19%	0.13%	99.80%
90	45.56%	25.24%	68.28%	17.73%	78.00%	14.83%	82.97%	13.33%	86.04%	12.42%	0.56%	99.22%
100	52.20%	22.03%	73.40%	15.79%	81.37%	13.51%	85.50%	12.33%	88.04%	11.62%	0.30%	99.41%

TABLE 6.20 The Detection Accuracy of *k*-Means Clustering Algorithm With/Without GATUD on SIRD

K	Without GATUD										With GATUD	
	$\theta = 0.01$		$\theta = 0.02$		$\theta = 0.03$		$\theta = 0.04$		$\theta = 0.05$			
	LER	F-M	LER	F-M	LER	F-M	LER	F-M	LER	F-M	LER	F-M
50	13.20%	80.09%	32.30%	62.74%	45.87%	52.44%	55.35%	45.76%	61.68%	41.40%	2.80%	**94.09%**
60	22.17%	71.64%	43.00%	53.86%	55.28%	45.25%	63.75%	39.72%	68.83%	36.34%	3.17%	**93.82%**
70	27.57%	64.89%	49.36%	48.09%	60.95%	40.33%	68.21%	35.97%	72.86%	33.23%	2.29%	**94.77%**
80	37.50%	56.39%	56.06%	43.16%	67.04%	36.51%	72.88%	32.95%	76.70%	30.72%	2.38%	**94.43%**
90	45.33%	49.58%	63.17%	38.59%	71.41%	33.45%	76.33%	30.51%	79.56%	28.77%	2.33%	**94.18%**
100	52.00%	42.35%	68.10%	34.20%	75.47%	30.16%	79.48%	28.01%	81.84%	26.71%	2.10%	**94.44%**

TABLE 6.21 The Detection Accuracy of _k_-Means Clustering Algorithm With/Without GATUD on SORD

K	Without GATUD										With GATUD	
	$\theta = 0.01$		$\theta = 0.02$		$\theta = 0.03$		$\theta = 0.04$		$\theta = 0.05$			
	LER	F-M	LER	F-M	LER	F-M	LER	F-M	LER	F-M	LER	F-M
50	18.00%	42.53%	36.40%	28.45%	50.27%	21.83%	60.60%	18.01%	67.52%	15.64%	2.00%	78.90%
60	22.67%	37.43%	44.50%	24.32%	58.83%	18.62%	68.04%	15.49%	73.50%	13.62%	1.67%	80.01%
70	33.00%	26.69%	54.86%	17.91%	66.81%	14.15%	73.75%	12.16%	78.06%	10.93%	2.29%	79.88%
80	43.00%	20.41%	63.69%	14.05%	73.58%	11.52%	79.16%	10.17%	82.53%	9.34%	1.50%	81.96%
90	48.67%	17.01%	68.61%	12.15%	77.63%	10.14%	82.19%	9.11%	85.04%	8.49%	1.44%	81.07%
100	58.40%	14.07%	74.20%	10.54%	81.03%	9.07%	84.75%	8.31%	87.04%	7.84%	1.50%	83.16%

TABLE 6.22 The Detection Accuracy of *k*-Means Clustering Algorithm With/Without GATUD on MORD

| | Without GATUD | | | | | | | | | | With GATUD | |
| | $\theta = 0.01$ | | $\theta = 0.02$ | | $\theta = 0.03$ | | $\theta = 0.04$ | | $\theta = 0.05$ | | | |
K	LER	F-M	LER	F-M	LER	F-M	LER	F-M	LER	F-M	LER	F-M
50	8.20%	72.61%	30.80%	48.03%	47.07%	36.94%	57.00%	31.08%	64.12%	27.23%	0.20%	**91.53%**
60	15.17%	56.72%	40.83%	37.64%	56.28%	29.59%	65.29%	25.16%	71.03%	22.42%	0.30%	**90.55%**
70	24.14%	44.26%	48.93%	30.09%	62.48%	24.24%	70.57%	20.99%	75.37%	19.04%	0.20%	**91.67%**
80	33.13%	37.11%	57.19%	25.44%	68.79%	20.90%	75.22%	18.52%	79.13%	17.05%	0.10%	**92.84%**
90	42.11%	29.29%	64.39%	21.18%	74.22%	17.97%	79.39%	16.27%	82.58%	15.26%	0.50%	**89.82%**
100	49.50%	24.53%	69.45%	18.61%	77.73%	16.16%	81.83%	14.91%	84.46%	14.15%	0.30%	**91.84%**

TABLE 6.23 The Detection Accuracy of *k*-Means Clustering Algorithm With/Without GATUD on MIRD

| | Without GATUD | | | | | | | | | | With GATUD | |
| | $\theta = 0.01$ | | $\theta = 0.02$ | | $\theta = 0.03$ | | $\theta = 0.04$ | | $\theta = 0.05$ | | | |
K	LER	F-M	LER	F-M	LER	F-M	LER	F-M	LER	F-M	LER	F-M
50	9.80%	84.02%	28.90%	61.31%	42.27%	49.73%	52.35%	42.47%	59.60%	37.69%	0.25%	99.31%
60	14.50%	75.38%	36.67%	53.65%	51.17%	43.10%	60.71%	36.99%	66.67%	33.19%	0.25%	99.41%
70	23.57%	63.42%	47.71%	44.59%	61.00%	36.16%	68.61%	31.70%	73.17%	28.97%	0.25%	99.51%
80	33.13%	53.07%	56.13%	38.25%	67.50%	31.78%	73.72%	28.35%	77.50%	26.25%	0.25%	99.70%
90	40.33%	45.60%	61.61%	33.55%	71.67%	28.50%	77.00%	25.84%	80.18%	24.25%	0.25%	99.11%
100	51.90%	37.75%	70.35%	28.72%	77.97%	25.19%	81.75%	23.38%	84.10%	22.28%	0.25%	99.70%

improve the accuracy of unsupervised intrusion detection techniques. This has been done by initially learning two labeled small data sets from the unlabeled data, where each data set represents either normal or abnormal behavior. Then a set of supervised classifiers are trained with question data sets to produce an ensemble-based decision-making model that can be integrated into both unsupervised anomaly scoring and clustering-based intrusion detection approaches, where, in the former, GATUD is used to mitigate the sensitivity of the anomaly threshold, while in the latter, it is used to efficiently label the produced clusters as either normal or abnormal. Experiments show that the integration of GATUD into SDAD described in Chapter 3, to mitigate the sensitivity of the anomaly threshold, has demonstrated significant results. Moreover, GATUD demonstrated significant and promising results when it was integrated into a clustering-based intrusion detection approach as a labeling method for the produced clusters.

CHAPTER 7

Threshold Password-Authenticated Secret Sharing Protocols

This chapter looks at designing appropriate authentication protocols for SCADA systems. Threshold password-authenticated secret sharing (TPASS) protocols allow a client to distribute a secret s amongst n servers and protect it with a password pw, so that the client can later recover the secret s from any subset of t of the servers using the password pw. This chapter presents two efficient TPASS protocols for SCADA systems, one that is built on a two-phase commitment and has a lower computation complexity and the other that is based on a zero-knowledge proof and has less communication rounds. Both protocols are particularly efficient for the client, who only needs to send a request and receive a response. Additionally, this chapter provides rigorous proofs of security for the protocols in the standard model. The experimental results have shown that the two TPASS protocols are more efficient than Camenisch et al.'s (2014) protocols and save up to 85–95% total computational time and up to 65–75% total communication overheads.

7.1 MOTIVATION

Threshold password-authenticated secret sharing (TPASS) protocols consider a scenario (Camenisch et al., 2014), inspired by the movie "Memento" in which the main character suffers from short-term memory loss, leads to an interesting cryptographic problem. Can a user securely recover his/her secrets from a set of servers (in a SCADA environment) if all the user can or wants to remember is a single password and all of the servers may be adversarial? In particular, can he/she protect his/her previous password when accidentally trying to run the recovery with all-malicious servers? A solution for this problem can act as a natural bridge from human-memorizable passwords to strong keys for cryptographic tasks.

A typical application of TPASS is to protect user data in Cloud or SCADA environments, where a client encrypts data with a random key before uploading it to a data server in the cloud, then secretly shares the random key together with a password with n key servers in the cloud. When he/she needs the key to decrypt the data downloaded from the data server, the client recovers the key from any subset of t of the n

SCADA Security: Machine Learning Concepts for Intrusion Detection and Prevention,
First Edition. Abdulmohsen Almalawi, Zahir Tari, Adil Fahad and Xun Yi.

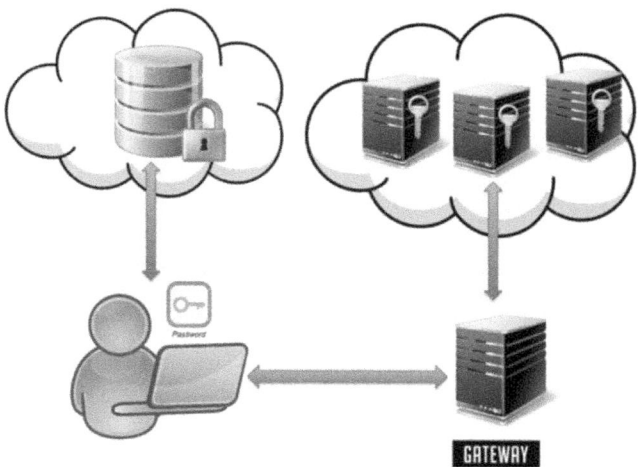

Figure 7.1   Security with TPASS.

key servers using the password. Therefore, to protect the data in the cloud/SCADA, the client needs to remember the password only. This process can be illustrated in Figure 7.1, where the gateway forwards messages between the client and key servers.

The first TPASS protocol was given by Bagherzandi et al. in 2011. It is built on the PKI model, secure under the decisional Diffie-Hellman assumption, using non-interactive zero-knowledge proofs. The basic idea is as follows: a client initially generates an ElGamal private and public key pairs $(sk, pk = g^{sk})$ (ElGamal, 1985) and secret-shares sk among servers using a t-out-of-n secret sharing (Shamir, 1979) and outputs public parameters including the public key pk and the encryptions $E(g^{\mathsf{pw}}, pk)$ and $E(s, pk)$ of password pw and secret s, respectively, under the public key pk. When retrieving the secret from the servers, the client encrypts the password $\mathsf{pw'}$ he remembers and sends the encryption $E(g^{\mathsf{pw'}}, pk)$ to the servers, each of which computes and returns $A_i = [E(g^{\mathsf{pw}}, pk)/E(g^{\mathsf{pw'}}, pk)]^{r_i} = E(g^{r_i(\mathsf{pw}-\mathsf{pw'})}, pk)$, where r_i is randomly chosen. The client then computes $A = \prod_{i=1}^{n} A_i$ and sends it to the servers. In the end, t servers cooperate to decrypt $B = E(s, pk)A = E(sg^{\sum r_i(\mathsf{pw}-\mathsf{pw'})}, pk)$ and sends partial decryptions to the client through secure channels, respectively. When $\mathsf{pw'} = \mathsf{pw}$, the client is able to retrieve the secret s by combining t partial decryptions. This protocol is secure against honest-but-curious adversaries but not malicious adversaries. A protocol against malicious adversaries was also given by Bagherzandi et al. (2011) using noninteractive zero-knowledge proofs. In this protocol, it is easy to see that the client must correctly remember the public key pk and the exact set of servers, as he or she sends out an encryption of his or her password attempt $\mathsf{pw'}$ he or she remembers. If pk can be tampered with and changed so that the adversary knows the decryption key, then the adversary can decrypt $\mathsf{pw'}$. Although the protocol actually encrypts $g^{\mathsf{pw'}}$, the malicious servers can perform an offline dictionary attack on $g^{\mathsf{pw'}}$ to obtain the password $\mathsf{pw'}$. In addition, even if the client can correctly remember the public key pk, the malicious servers can cheat the client with a different secret

s' by sending the partial decryptions of $B' = E(s', pk)A$ instead of $B = E(s, pk)A$ to the client. A 1-out-of-2 TPASS was given by Camenisch et al. in 2012. This protocol also leaks the password when the client tries to retrieve his or her secret from a set of all-malicious servers.

Authenticating to the wrong servers is a common scenario when users are tricked in phishing attacks. To overcome this shortcoming, Camenisch et al. (2014) used the first t-out-of-n TPASS protocol for any $n > t$. The protocol requires the client to only remember a username and a password, assuming that a PKI is available. If the client misremembers his or her list of servers and tries to retrieve his or her secret from corrupt servers, the protocol prevents the servers from learning anything about the password or secret, as well as from planting a different secret into the user's mind than the secret that he or she stored earlier. The construction of the Camenisch et al. protocol is inspired by the Bagherzandi et al. protocol based on a homomorphic threshold encryption scheme, but the crucial difference is that in the retrieval protocol of Camenisch et al. the client never sends out an encryption of his or her password attempt. Instead, the client derives an encryption of the (randomized) quotient of the password used at setup and the password attempt. The servers then jointly decrypt the quotient and verify whether it yields "1", indicating that both passwords matched. In case the passwords were not the same, all the servers learn is a random value.

The Camenisch et al. TPASS protocol (2014) proved to be secure in the UC framework and requires the client to be involved in many communication rounds so that it becomes impractical for the client. The client has to do $5n + 15$ exponentiations in $\mathbb{G}$ for the setup protocol and $14t + 24$ exponentiations in the retrieval protocol. Each server has to perform $n + 18$ and $7t + 28$ exponentiations in these respective protocols.

Recently, Abdalla et al. (2016) used new new robust password-protected secret sharing protocols that are significantly more efficient than the existing ones. Their protocols have been proven in the random-oracle model, because their construction requires random nonmalleable fingerprints, which are provided by an ideal hash function. In addition, Jarecki et al. (2017) constructed password-protected secret sharing protocols based on oblivious pseudorandom functions, formulated as a universally composable (UC) functionality.

To improve the efficiency of TPASS, this chapter proposes two new t-out-of-n TPASS protocols for any $n > t$: one protocol is built on a two-phase commitment (and has lower computation complexity, useful for SCADA and Cloud environments) and another protocol is built on zero-knowledge proof and has less communication rounds. Both protocols are in particular efficient for the client, who only needs to send a request and receive a response. The basic idea is as follows: a client initially secret-shares a password, a secret, and the digest of the secret with n servers, such as that t out of the n servers can recover the secret. When retrieving the secret from the servers, the client submits to the servers $A = g_1^r g_2^{\mathrm{pw}_C}$, where r is randomly chosen and pw_C is the password, and then t servers cooperate to generate and return an ElGamal encryption of the secret and an ElGamal encryption of the digest of the secret, both under the public key g_1^r. Two-phase commitment and zero-knowledge proof are used to prevent the collusion attack from up to $t - 1$ malicious servers. At the end, the client then decrypts the two ciphertexts and accepts the secret if one decrypted value is another's digest.

The protocols described in this chapter are significantly more efficient than the Camenisch et al. (2014) protocol in terms of computational and communication overheads. Indeed, the client only needs to send a request and receive a response. In addition, the client needs to do $3n$ evaluations of polynomials of degree $t - 1$ in $\mathbb{Z}_q$, Yifor the initialization, and seven exponentiations for the retrieval protocol. Each server only needs to do $t + 10$ (in the two-phase commitment case) or $3t + 10$ (in the one-phase zero-knowledge proof case) exponentiations in the retrieval protocol. The computation and communication complexities for the client are independent of the number of servers n and the threshold t.

A rigorous proof of security for the protocols in the standard model is described here. Like the Camenisch et al. (2014) protocol, the protocols can protect the password of the client even if he/she communicates with all-malicious servers by mistake. In addition, they prevent the servers from planting a different secret into the user's mind than the secret that was stored earlier. This chapter is an extended version of the paper that appeared in Yi et al. (2015), where a new TPASS protocol based on a zero-knowledge proof is addeded, which reduces the communications among servers from two phases to one phase. In this way, these new protocols could be easy and efficiently used in critical systems, such as SCADA.

7.2 EXISTING SOLUTIONS

A close work related to TPASS is the threshold password – an authenticated key exchange (TPAKE), which lets the client agree on a fresh session key with each of the servers, but does not allow the client to store and recover a secret. Depending on the desired security properties, one can build a TPASS scheme from a TPAKE scheme by using the agreed-upon session keys to transmit the stored secret shares over secure channels (Bagherzandi et al., 2011).

The first TPAKE protocols, due to Ford and Kaliski (2000) and Jablon (2001), were not proved to be secure. The first provably secure TPAKE protocol, a t-out-of-n protocol in a PKI setting, was by MacKenzie et al. MacKenzie et al. (2006). The 1-out-of-2 protocol of Brainard et al. (2003) is implemented in EMC's RSA Distributed Credential Protection. Both protocols either leak the password or allow an offline dictionary attack when the retrieval is performed with corrupt servers. The t-out-of-n TPAKE protocols by Di Raimondo and Gennaro (2006) and the 1-out-of-2 protocol by Katz et al. (2005) are proved secure in a hybrid password-only/PKI setting, where the user does not know any public keys, but the servers and an intermediate gateway do have a PKI. These protocols actually remain secure when executed with all-corrupt servers, but are restricted to the cases that $n > 3t$ and $(t, n) = (1, 2)$. Based on identity-based encryption (IBE), a 1-out-of-2 protocol, where the client is required to remember the identities of the two servers besides his or her password, was by Yi et al. (2014).

7.3 DEFINITION OF SECURITY

This section defines the security for the TPASS protocol on the basis of the security models for PAKE (Bellare et al., 2000; Katz et al., 2001).

Participants, Initialization, Passwords, Secrets. A TPASS protocol involves three kinds of protocol participants: (1) a group of clients (denoted as Client), each of which requests TPASS services from t servers on the network; (2) a group of n servers $S_1, S_2, \ldots, S_n$ (denoted as Server $= \{S_1, S_2, \ldots, S_n\}$), which cooperate to provide TPASS services to clients on the network; (3) a gateway (GW), which coordinates TPASS. We assume that User $=$ Client $\bigcup$ Server and Client $\bigcap$ Server $= \emptyset$. When the gateway GW coordinates TPASS, it simply forwards messages between a client and t servers.

Prior to any execution of the protocol, we assume that an initialization phase occurs. During initialization, the n servers cooperate to generate public parameters for the protocol, which are available to all participants.

It is also assumed that the client C chooses its password pw_C independently and uniformly at random from a "dictionary" $D = \{pw_1, pw_2, \ldots, pw_N\}$ of size N, where N is a fixed constant that is independent of any security parameter. The client then secretly shares the password with the n servers such that any t servers can restore the password.

In addition, the client C is also assumed to choose the secret s_C independently and uniformly at random from $\mathbb{Z}_q^*$, where q is a public parameter. The client then secretly shares the secret with the n servers such that any t servers can recover the secret.

We assume that at least $n - t + 1$ servers are trusted not to collude to determine the password and the secret of the client. The client C needs to remember pw_C only to retrieve its secret s_C.

Execution of the Protocol. A protocol determines how users behave in response to input from their environments. In the formal model, these inputs are provided by the adversary. Each user is assumed to be able to execute the protocol multiple times (possibly concurrently) with different partners. This is modeled by allowing each user to have an unlimited number of instances with which to execute the protocol. We denote instance i of user U as U^i. A given instance may be used only once. The adversary is given oracle access to these different instances. Furthermore, each instance maintains a (local) state, which is updated during the course of the experiment. In particular, each instance U^i is associated with the following variables, initialized as NULL or FALSE (as appropriate) during the initialization phase.

- sid_U^i is a variable containing the session identity for an instance U^i. The session identity is simply a way to keep track of the different executions of a particular user U. Without loss of generality, we simply let this be the (ordered) concatenation of all messages sent and received by instance U^i.
- s_C^i is a variable containing the secret s_C for a client instance C^i. Retrieval of the secret is, of course, the ultimate goal of the protocol.

- acc^i_U and $term^i_U$ are boolean variables denoting whether a given instance U^i has been accepted or terminated, respectively. Termination means that the given instance has done receiving and sending messages, acceptance indicates successful termination. When an instance U^i has been accepted, sid^i_U is no longer NULL. When a client instance C^i has been accepted, s^i_C is no longer NULL.
- $state^i_U$ records any state necessary for execution of the protocol by U^i.
- $used^i_U$ is a boolean variable denoting whether an instance U^i has begun executing the protocol. This is a formalism that will ensure each instance is used only once.

The adversary $\mathcal{A}$ is assumed to have complete control over all communications in the network (between the clients and servers and between the servers and servers) and the adversary's interaction with the users (more specifically, with various instances) is modeled via access to oracles. The state of an instance may be updated during an oracle call and the oracle's output may depend upon the relevant instance. The oracle types include:

- Send(C, i, M) – This sends message M to a client instance C^i. Assuming $term^i_C = $ FALSE, this instance runs according to the protocol specification, updating the the state as appropriate. The output of C^i (i.e., the message sent by the instance) is given to the adversary, who receives the updated values of sid^i_C, acc^i_C, and $term^i_C$. This oracle call models an active attack to the protocol. If M is empty, this query represents a prompt for C to initiate the protocol.
- Send(S, j, U, M) – This sends message M to a server instance S^j, supposedly from a user U (either a client or a server) or even a set of servers. Assuming $term^j_S = $ FALSE, this instance runs according to the protocol specification, updating state as appropriate. The output of S^j (i.e., the message sent by the instance) is given to the adversary, who receives the updated values of sid^j_S, acc^j_S, and $term^j_S$. If S is corrupted, the adversary also receives the entire internal state of S. This oracle call also models an active attack to the protocol.
- Execute($C, i, \mathbb{S}$) – If the client instance C^i and t server instances, denoted as $\mathbb{S}$, have not yet been used, this oracle executes the protocol between these instances and outputs the transcript of this execution. This oracle call represents passive eavesdropping of a protocol execution. In addition to the transcript, the adversary receives the values of sid, acc, and $term$ for client and server instances, at each step of protocol execution. In addition, if any server in $\mathbb{S}$ is corrupted, the adversary is given the entire internal state of the server.
- Corrupt(S) – This sends the password and secret shares of all clients stored in the server S to the adversary. This oracle models possible compromising of a server due to, for example, hacking into the server.
- Corrupt(C) – This query allows the adversary to learn the password of the client C and then the secret of the client, which models the possibility of subverting a client by, for example, witnessing a user typing in his password or installing a "Trojan horse" on his machine.

- Test(C, i) – This oracle does not model any real-world capability of the adversary, but is instead used to define security. If $acc_C^i = $ TRUE, a random bit b is generated. If $b = 0$, the adversary is given s_C^i and if $b = 1$ the adversary is given a random number. The adversary is allowed only a single *Test* query, at any time during its execution.

Correctness. To be viable, a TPASS protocol must satisfy the following notion of correctness: If a client instance C^i and t server instances $\mathbb{S}$ run an honest execution of the protocol with no interference from the adversary, then $acc_C^i = acc_S^j = $ TRUE for any server instance S^j in $\mathbb{S}$.

Freshness. To formally define the adversary's success we need to define a notion of freshness for a client, where freshness of the client is meant to indicate that the adversary does not trivially know the value of the secret of the client. We say a client instance C^i is fresh if (1) C has not been corrupted; (2) Test(C) has not been queried; and (3) at least $n - t + 1$ out of n servers are not corrupted.

Advantage of the Adversary. We consider passive and active attacks, respectively. In a passive attack, the adversary is allowed to call Execute, Corrupt, and Testt oracles. Informally, a passive adversary succeeds if it can guess the bit b used by the Test oracle. We say a passive adversary $\mathcal{A}$ succeeds if it makes a query Test(C, i) to a fresh client instance C^i, with $acc_C^i = $ TRUE at the time of this query, and outputs a bit b' with $b' = b$ (recall that b is the bit chosen by the Test oracle). We denote this event by Succ_P. The advantage of a passive adversary $\mathcal{A}$ in attacking protocol P is then given by

$$\mathsf{Adv}_{P\mathcal{A}}^P(k) = 2 \cdot \Pr[\mathsf{Succ}_P] - 1$$

where the probability is taken over the random coins used by the adversary and the random coins used during the course of the experiment (including the initialization phase).

Definition 7.1 Protocol P is a secure TPASS protocol against the passive attack, if, for all passive PPT adversaries $\mathcal{A}$, there exists a negligible function $\varepsilon(\cdot)$ such that for a security parameter k,

$$\mathsf{Adv}_{P\mathcal{A}}^P(k) \leq \varepsilon(k)$$

In an active attack, the adversary is allowed to call Send and Corrupt oracles. Informally, an active adversary succeeds if it can convince a client to accept a wrong secret key. We say an active adversary $\mathcal{A}$ succeeds if it makes a query Send(C, i) to a fresh client instance C^i, resulting in $acc_C^i = $ TRUE. We denote this event by Succ_A. The advantage of an active adversary $\mathcal{A}$ in attacking protocol P is then given by

$$\mathsf{Adv}_{A\mathcal{A}}^P(k) = \Pr[\mathsf{Succ}_A]$$

where the probability is taken over the random coins used by the adversary and the random coins used during the course of the experiment (including the initialization phase).

The active adversary can always succeed by trying all passwords one by one in an on-line impersonation attack. A protocol is secure against the active attack if this is the best an adversary can do. The on-line attacks correspond to Send queries. Formally, each instance for which the adversary has made a Send query counts as one on-line attack. The number of on-line attacks represents a bound on the number of passwords the adversary could have tested in an on-line fashion.

Definition 7.2 Protocol P is a secure TPASS protocol against the active attack if, for all dictionary size N and for all active PPT adversaries $\mathcal{A}$ making at most $Q(k)$ on-line attacks, there exists a negligible function $\varepsilon(\cdot)$ such that for a security parameter k,

$$\mathsf{Adv}^P_{\mathcal{A},\mathcal{A}}(k) \le Q(k)/N + \varepsilon(k)$$

7.4 TPASS PROTOCOLS

This section describes the two TPASS protocols based on the two-phase commitment protocol and zero-knowledge proof, respectively.

7.4.1 Protocol-Based on Two-Phase Commitment

Initialization. Given a security parameter $k \in z^*$, the initialization includes:

Parameter Generation. On input k, the n servers agree on a cyclic group $\mathbb{G}$ of large prime order q with a generator g_1 and a hash function $H : \{0,1\}^* \to \mathbb{Z}_q$. Then the n servers cooperate to generate g_2 like Yi et al. (2013), such that no one knows the discrete logarithm of g_2 based on g_1 if one out of the n server is honest. The public parameters for the protocol is params $= \{\mathbb{G}, q, g_1, g_2, H\}$.

Password Generation. On input params, each client $C \in$ Client with identity ID_C uniformly draws a string pw_C, the password, from the dictionary $D = \{\mathsf{pw}_1, \mathsf{pw}_2, \dots, \mathsf{pw}_N\}$. The client then randomly chooses a polynomial $f_1(x)$ of degree $t-1$ over $\mathbb{Z}_q$ such that $\mathsf{pw}_C = f_1(0)$, and distributes $\{ID_C, i, f_1(i)\}$ to the server S_i via a secure channel, where $i = 1, 2, \dots, n$.

Secret Sharing. On input params, each client $C \in$ Client randomly chooses s from $\mathbb{Z}_q^*$. The client then randomly chooses two polynomials $f_2(x)$ and $f_3(x)$ of degree $t-1$ over $\mathbb{Z}_q$ such that $s = f_2(0)$ and $H(g_2^s) = f_3(0)$, and distributes $\{ID_C, i, f_2(i), f_3(i)\}$ to the server S_i via a secure channel, where $i = 1, 2, \dots, n$. We define the secret s_C as g_2^s.

Protocol Execution. Given the public params $= \{\mathbb{G}, q, g_1, g_2, H\}$, the client C (knowing its identity ID_C and password pw_C) runs TPASS protocol P with t servers (each server knowing $\{ID_C, i, f_1(i), f_2(i), f_3(i)\}$) to retrieve the secret s_C. As shown in Figure 7.2, TPASS protocol is executed with three algorithms as follows.

(A) Retrieval Request. Given the public parameters $\{\mathbb{G}, g_1, g_2, q, H\}$, the client C with the identity ID_C validates if q is a large prime and $g_1^q = g_2^q = 1$. If so, the client, who remembers the password pw_C, randomly chooses r from $\mathbb{Z}_q^*$ and computes

$$A = g_1^r g_2^{-\mathsf{pw}_C}.$$

Then the client submits $\mathsf{msg}_C = \langle ID_C, A \rangle$ to the gateway GW for the n servers.
Remark. The purpose for the client to validate the public parameters is to ensure that the discrete logarithm over $\{\mathbb{G}, q, g_1, g_2\}$ is hard in case the adversary can change the public parameters.

(B) Retrieval Response. After receiving the request msg_C from the client C, the gateway GW forwards it to t available servers in response in the request. Without loss of generality, we assume that the first t servers, denoted as $\mathbb{S} = \{S_1, S_2, \ldots, S_t\}$, cooperate to generate a response. There are two phases for the t servers to generate a retrieval response.

Commitment Phase. Based on the identity ID_C of the client, each server S_i ($i = 1, 2, \ldots, t$) randomly chooses r_i, c_i, d_i from $\mathbb{Z}_q^*$ and computes

$$B_i = g_1^{r_i} g_2^{a_i f_1(i)}, C_i = g_1^{c_i}, D_i = g_1^{d_i}, \delta_i = g_1^{H(ID_C, A, B_i, C_i, D_i)}$$

where $a_i = \prod_{1 \leq j \leq t, j \neq i} \frac{j}{j-i}$.
In the commitment phase, S_i broadcasts its commitment $\langle ID_C, \delta_i \rangle$.

Opening Phase. After receiving all commitments $\langle ID_C, \delta_j \rangle$ $(1 \leq j \leq t)$, S_i broadcasts its opening $\langle ID_C, B_i, C_i, D_i \rangle$.
Each server S_i verifies if $\delta_j = g_1^{H(ID_C, A, B_j, C_j, D_j)}$ for all $j \neq i$. If so, based on the identity ID_C of the client, S_i computes

$$C = \prod_{j=1}^{t} C_j, D = \prod_{j=1}^{t} D_j, h_i = H(ID_C, A, C, D)$$

$$E_i = g_2^{a_i f_2(i) h_i} C^{-r_i} (A \prod_{j=1}^{t} B_j)^{c_i}, F_i = g_2^{a_i f_3(i) h_i} D^{-r_i} (A \prod_{j=1}^{t} B_j)^{d_i}$$

and sets $\mathsf{acc}_{S_i} = \mathsf{TRUE}$.
Then S_i sends $\mathsf{msg}_i^* = \{ID_C, C, D, E_i, F_i\}$ to the gateway GW.

Public: $\mathbb{G}, q, g_1, g_2, H$

Client C
ID_C, pw_C

$r \xleftarrow{R} \mathbb{Z}_q^*$
$A = g_1^r g_2^{-\mathsf{pw}_C}$

Server S_i

$\{ID_C, i, f_1(i), f_2(i), f_3(i)\}$
$i = 1, 2, \cdots, t$

$\xrightarrow{\mathsf{msg}_C = \langle ID_C, A \rangle}$ Gateway $\xrightarrow{\mathsf{msg}_C = \langle ID_C, A \rangle}$
GW $\mathbb{S} = \{\mathsf{S}_1, \mathsf{S}_2, \cdots, \mathsf{S}_t\}$

$r_i, c_i, d_i \xleftarrow{R} \mathbb{Z}_q^*, a_i = \prod_{1 \le j \le t, j \ne i} \frac{j}{j-i}$
$B_i = g_1^{r_i} g_2^{a_i f_1(i)}$
$C_i = g_1^{c_i}, D_i = g_1^{d_i}$
$\delta_i = g_1^{H(ID_C, A, B_i, C_i, D_i)}$
$\mathsf{msg}_i = \langle ID_C, \delta_i, B_i, C_i, D_i \rangle$

$\xrightarrow{\langle ID_C, \delta_i \rangle, \ i = 1, 2, \cdots, t}$ Phase 1
S_i broadcasts commit in $\mathbb{S}$

$\xrightarrow{\langle ID_C, B_i, C_i, D_i \rangle, \ i = 1, 2, \cdots, t}$ Phase 2
S_i broadcasts opening in $\mathbb{S}$

if $\delta_j = g_1^{H(ID_c, A, B_j, C_j, D_j)}$ $(1 \le j \le t)$
$C = \prod_{j=1}^t C_j, D = \prod_{j=1}^t D_j$
$h_i = H(ID_C, A, C, D)$
$E_i = g_2^{a_i f_2(i) h_i} C^{-r_i} (A \prod_{j=1}^t B_j)^{c_i}$
$F_i = g_2^{a_i f_3(i) h_i} D^{-r_i} (A \prod_{j=1}^t B_j)^{d_i}$
$\mathsf{acc}_{\mathsf{S}_i} = \mathsf{TRUE}$
else return $\perp$

Gateway $\xleftarrow{\mathsf{msg}_i^* = \langle ID_C, C, D, E_i, F_i \rangle}$
GW
$E = \prod_{i=1}^t E_i$
$F = \prod_{i=1}^t F_i$

$\xleftarrow{\mathsf{msg}_S = \langle ID_C, C, D, E, F \rangle}$

$h = H(ID_C, A, C, D)$
$S = (E/C^r)^{h^{-1}}$
$T = (F/D^r)^{h^{-1}}$
if $T = g_2^{H(S)}$, $\mathsf{acc}_C = \mathsf{TRUE}$
else return $\perp$

Figure 7.2 The TPASS Protocol P based on a Two-Phase Commitment.

The gateway GW computes

$$E = \prod_{i=1}^{t} E_i, F = \prod_{i=1}^{t} F_i$$

and returns to the client with $\text{msg}_S = \{ID_C, C, D, E, F\}$.

(C) Secret Retrieval. After receiving the response $\text{msg}_S = \{ID_C, C, D, E, F\}$ from the gateway, the client computes

$$h = H(ID_C, A, C, D), S = (E/C^r)^{h^{-1}}, T = (F/D^r)^{h^{-1}}$$

and verifies if $T = g_2^{H(S)}$. If so, the client sets $\text{acc}_C = \text{TRUE}$, the secret $s_C = S$, and $\perp$ otherwise.

7.4.2 Protocol Based on Zero-Knowledge Proof

Initialization. The initialization is the same as described in Section 7.4.1.

Protocol Execution. Given the public $\text{params} = \{\mathbb{G}, q, g_1, g_2, H\}$, the client C (knowing its identity ID_C and password pw_C) runs TPASS protocol P with t servers (each server knowing $\{ID, i, f_1(i), f_2(i), f_3(i)\}$) to retrieve the secret s_C. As shown in Figure 7.3, TPASS protocol is executed with three algorithms as follows.

(A) Retrieval Request. Given the public parameters $\{\mathbb{G}, g_1, g_2, q, H\}$, the client C with the identity ID_C validates if q is a large prime and $g_1^q = g_2^q = 1$. If so, the client, who remembers the password pw_C, randomly chooses r from $\mathbb{Z}_q^*$ and computes

$$A = g_1^r g_2^{-\text{pw}_C}$$

Then the client submits $\text{msg}_C = \langle ID_C, A \rangle$ to the gateway GW for the n servers.

(B) Retrieval Response. After receiving the request msg_C from the client C, the gateway GW forwards it to t available servers to respond the request. Without loss of generality, we assume that the first t servers, denoted as $\mathbb{S} = \{S_1, S_2, \cdots, S_t\}$, cooperate to generate a response. Unlike the protocol based on two-phase commitment, the t servers can generate the response in one phase.

Based on the identity ID_C of the client, each server S_i ($i = 1, 2, \ldots, t$) randomly chooses r_i, c_i, d_i from $\mathbb{Z}_q^*$ and computes

$$B_i = g_1^{r_i} g_2^{a_i f_1(i)}, C_i = g_1^{c_i}, D_i = g_1^{d_i}$$

$$h_i = H(ID_C, A, B_i, C_i, D_i), H_i = H(h_i), \delta_i = h_i c_i + H_i d_i \pmod{q}$$

Public: $\mathbb{G}, q, g_1, g_2, H$

Client C   **Server S_i**

ID_C, pw_C

$$r \overset{R}{\leftarrow} \mathbb{Z}_q^*$$
$$A = g_1^r g_2^{-\mathsf{pw}_C}$$

$\qquad\qquad\{ID_C, i, f_1(i), f_2(i), f_3(i)\}$
$\qquad\qquad\qquad i = 1, 2, \cdots, t$

$\overset{\mathsf{msg}_C = \langle ID_C, A\rangle}{\longrightarrow}$ Gateway $\overset{\mathsf{msg}_C = \langle ID_C, A\rangle}{\longrightarrow}$

GW $\quad \mathbb{S} = \{S_1, S_2, \cdots, S_t\}$

$$r_i, c_i, d_i \overset{R}{\leftarrow} \mathbb{Z}_q^*, a_i = \prod_{1 \le j \le t, j \ne i} \frac{j}{j-i}$$
$$B_i = g_1^{r_i} g_2^{a_i f_1(i)}$$
$$C_i = g_1^{c_i}, D_i = g_1^{d_i}$$
$$h_i = H(ID_C, A, B_i, C_i, D_i), H_i = H(h_i)$$
$$\delta_i = h_i c_i + H_i d_i (mod\ q) \text{ (ZKP)}$$

$\overset{\mathsf{msg}_i = \langle ID_C, B_i, C_i, D_i, \delta_i\rangle}{\longrightarrow}$

S_i broadcasts msg_i in $\mathbb{S}$

for $j = 1$ to t where $j \ne i$
$\{h_j = H(ID_C, A, B_j, C_j, D_j), H_j = H(h_j)\}$
if $g_1^{\delta_j} = C_j^{h_j} D_j^{H_j}$ $(1 \le j \le t$ where $j \ne i)$
$\{C = \prod_{j=1}^t C_j, D = \prod_{j=1}^t D_j$
$h = H(ID_C, A, C, D)$
$E_i = g_2^{a_i f_2(i) h} C^{-r_i} (A \prod_{j=1}^t B_j)^{c_i}$
$F_i = g_2^{a_i f_3(i) h} D^{-r_i} (A \prod_{j=1}^t B_j)^{d_i}$
$\mathsf{acc}_{S_i} = \mathsf{TRUE}\}$
else return $\perp$

Gateway $\overset{\mathsf{msg}_i^* = \langle ID_C, C, D, E_i, F_i\rangle}{\longleftarrow}$
GW

$$E = \prod_{i=1}^t E_i$$
$$F = \prod_{i=1}^t F_i$$

$\overset{\mathsf{msg}_S = \langle ID_C, C, D, E, F\rangle}{\longleftarrow}$

$h = H(ID_C, A, C, D)$
$S = (E/C^r)^{h^{-1}}$
$T = (F/D^r)^{h^{-1}}$
if $T = g_2^{H(S)}$, $\mathsf{acc}_C = \mathsf{TRUE}$
else return $\perp$

Figure 7.3   The TPASS Protocol P based on Zero-Knowledge Proof.

where $a_i = \prod_{1\leq j\leq t, j\neq i}\frac{j}{j-i}$, (C_i, D_i, δ_i) is a zero-knowledge proof of knowledge of (c_i, d_i).

Then S_i broadcasts $\mathsf{msg}_i = \langle ID_C, B_i, C_i, D_i, \delta_i\rangle$ in $\mathbb{S}$.

Each server S_i computes

$$h_j = H(ID_C, A, B_j, C_j, D_j), H_j = H(h_j)$$

for all $j\neq i$ and verifies the zero-knowledge proof of knowledge of (c_j, d_j) by checking if

$$g_1^{\delta_j} = C_j^{h_j} D_j^{H_j}$$

for all $j\neq i$. If so, based on the identity ID_C of the client, S_i computes

$$C = \prod_{j=1}^t C_j, D = \prod_{j=1}^t D_j, h = H(ID_C, A, C, D)$$

$$E_i = g_2^{a_i f_2(i)h} C^{-r_i}(A\prod_{j=1}^t B_j)^{c_i}, F_i = g_2^{a_i f_3(i)h} D^{-r_i}(A\prod_{j=1}^t B_j)^{d_i}$$

and sets $acc_{S_i} = TRUE$.

Then S_i sends $\mathsf{msg}_i^* = \{ID_C, C, D, E_i, F_i\}$ to the gateway GW.

The gateway GW computes

$$E = \prod_{i=1}^t E_i, F = \prod_{i-1}^t F_i$$

and returns to the client with $\mathsf{msg}_S = \{ID_C, C, D, E, F\}$.

(C) Secret Retrieval. After receiving the response $\mathsf{msg}_S = \{ID_C, C, D, E, F\}$ from the gateway, the client computes

$$h = H(ID_C, A, C, D), S = (E/C^r)^{h^{-1}}, T = (F/D^r)^{h^{-1}}$$

and verifies if $T = g_2^{H(S)}$. If so, the client sets $acc_C = TRUE$, the secret $s_C = S$, and $\perp$ otherwise.

7.4.3 Correctness

Correctness of the TPASS Protocol Based on Two-Phase Commitment.
Let us assume that a client instance C^i and t server instances $\mathbb{S}$ run an honest execution

of the TPASS protocol P with no interference from the adversary. With reference to Figure 7.2, it is obvious that $acc_{S_j} = TRUE$ for $1 \leq j \leq t$. In addition, we have

$$C = \prod_{j=1}^{t} C_j = g_1^{\sum_{j=1}^{t} c_j}, D = \prod_{j=1}^{t} D_j = g_1^{\sum_{j=1}^{t} d_j}$$

$$E_i = g_2^{a_i f_2(i) h_i} C^{-r_i} (A \prod_{j=1}^{t} B_j)^{c_i}$$

$$= g_2^{a_i f_2(i) h_i} g_1^{-r_i \sum_{j=1}^{t} c_j} (g_1^r g_2^{-\mathsf{pw}c} g_1^{\sum_{j=1}^{t} r_j} g_2^{\mathsf{pw}c})^{c_i}$$

$$= g_2^{a_i f_2(i) h_i} g_1^{-r_i \sum_{j=1}^{t} c_j} g_1^{c_i \sum_{j=1}^{t} r_j} g_1^{c_i r}$$

$$F_i = g_2^{a_i f_3(i) h_i} D^{-r_i} (A \prod_{j=1}^{t} B_j)^{d_i}$$

$$= g_2^{a_i f_3(i) h_i} g_1^{-r_i \sum_{j=1}^{t} d_j} (g_1^r g_2^{-\mathsf{pw}c} g_1^{\sum_{j=1}^{t} r_j} g_2^{\mathsf{pw}c})^{d_i}$$

$$= g_2^{a_i f_3(i) h_i} g_1^{-r_i \sum_{j=1}^{t} d_j} g_1^{d_i \sum_{j=1}^{t} r_j} g_1^{d_i r}$$

$$h = H(ID_C, A, C, D)$$

$$E = \prod_{i=1}^{t} E_i = \prod_{i=1}^{t} g_2^{a_i f_2(i) h} g_1^{-r_i \sum_{j=1}^{t} c_j} g_1^{c_i \sum_{j=1}^{t} r_j} g_1^{c_i r}$$

$$= g_2^{sh} g_1^{-\sum_{i=1}^{t} r_i \sum_{j=1}^{t} c_j} g_1^{\sum_{i=1}^{t} c_i \sum_{j=1}^{t} r_j} g_1^{r \sum_{i=1}^{t} c_i} = g_2^{sh} C^r$$

$$F = \prod_{i=1}^{t} F_i = \prod_{i=1}^{t} g_2^{a_i f_3(i) h} g_1^{-r_i \sum_{j=1}^{t} d_j} g_1^{d_i \sum_{j=1}^{t} r_j} g_1^{d_i r}$$

$$= g_2^{H(g_2^s)h} g_1^{-\sum_{i=1}^{t} r_i \sum_{j=1}^{t} d_j} g_1^{\sum_{i=1}^{t} d_i \sum_{j=1}^{t} r_j} g_1^{r \sum_{i=1}^{t} d_i} = g_2^{H(g_2^s)h} D^r$$

We can see that (C, E) and (D, F) are in fact the EGamal encryptions of g_2^{sh} and $g_2^{H(g_2^s)h}$ under the public key g_1^r, respectively. Therefore, we have $acc_C = TRUE$ because

$$h = H(ID_C, A, C, D)$$

$$S = (E/C^r)^{h^{-1}} = (g_2^{sh})^{h^{-1}} = g_2^s$$

$$T = (F/D^r)^{h^{-1}} = (g_2^{H(g_2^s)h})^{h^{-1}} = g_2^{H(g_2^s)}$$
$$T = g_2^{H(S)}$$

In summary, the TPASS protocol based on two-phase commitment has correctness.

Correctness of the TPASS Protocol Based on Zero-Knowledge Proof.
Comparing the two protocols shown in Figures 7.2 and 7.3, we can see that there are two differences: (1) δ_i is computed differently; (2) δ_i is verified differently. Therefore, we only need to show that $g_1^{\delta_j} = C_j^{h_j} D_j^{H_j}$ for $j = 1, 2, \ldots, t$ for the second protocol if each server S_i follows the protocol to compute B_i, C_i, D_i, and δ_i.

Because $C_j = g_1^{c_j}, D_j = g_1^{d_j}, \delta_j = h_j c_j + H_j d_j$ for $j = 1, 2, \ldots, t$, we have

$$g_1^{\delta_j} = g_1^{h_j c_j + H_j d_j} = (g_1^{c_j})^{h_j}(g_1^{d_j})^{H_j} = C_j^{h_j} D_j^{H_j}$$

for $j = 1, 2, \ldots, t$. Therefore, the TPASS protocol based on zero-knowledge proof has correctness.

7.4.4 Efficiency

Efficiency of the TPASS Protocol Based on Two-Phase Commitment. In the first protocol, the client needs to compute seven exponentiations in $\mathbb{G}$ and send or receive five group elements in $\mathbb{G}$. Each server needs to compute $t + 10$ exponentiations in $\mathbb{G}$ and send or receive $4t + 5$ group elements in $\mathbb{G}$.

The client involves only two communication rounds with the gateway, i.e., sending msg_C to the gateway and receiving msg_S from the gateway. Each server S_i participates in six communication rounds with other servers and the gateway, i.e., receiving msg_C from the gateway, broadcasting the commitment $\langle ID_C, \delta_i \rangle$ to other servers, receiving $\langle ID_C, \delta_j \rangle$ for all $j \neq i$ from other servers, broadcasting $\langle ID_C, B_i, C_i, D_i \rangle$, receiving $\langle ID_C, B_j, C_j, D_j \rangle$ for all $j \neq i$, and finally sending msg_i^* to the gateway.

The performance comparison of Camenisch et al. (2014) protocol (with provable security in the UC framework) and the protocol (with provable security in the standard model) is shown in Table 7.1. In this table, exp. represents the computation complexity of a modular exponentiation, $|g|$ is the size of a group element in $\mathbb{G}$, and $|q|$ is the size of a group element in $\mathbb{Z}_q$, while C, S, and G stand for Client, Server and Gateway, respectively. In addition, $pk = \prod epk_i, tpk = \prod tpk_i$. In the Camenisch et al. (2014) protocol, a hash value is counted as half a group element.

In the initialization phase, the client secret-shares the password, secret, and the digest of the secret with the n servers via n secure channels, which may be established with PKI. In the setup protocol of Camenisch et al. (2014), the client sets up the shares with the n servers based on PKI. The retrieval protocol does not rely on PKI, but the retrieval protocol of Camenisch et al. still requires PKI. In view of this, the retrieval protocol can be implemented easier than the Camenisch et al. retrieval protocol.

TABLE 7.1 Performance Comparison of the Camenisch et al. Protocol and the described Protocols.

	Camenisch et al. (Camenisch et al., 2014)	Protocol 1	Protocol 2
Public Keys	C: username $S{:}S_i{:}epk_i, spk_i, tpk_i$	C: username SS_i: none	
Private Keys	$C{:}\mathsf{pw}_C$ $S{:}S_i{:}esk_i, ssk_i, tsk_i$ $E(\mathsf{pw}_C, pk), E(s, pk)$ $E(\mathsf{pw}_C, tpk), E(s, tpk)$	$C{:}\mathsf{pw}_C$ $S{:}S_i{:}f_1(i), f_2(i), f_3(i)$ where $\sum a_i f_1(i) = \mathsf{pw}_C, \sum a_i f_2(i) = s$ and $\sum a_i f_3(i) = H(g_2^s)$	
Setup Comp. Complexity	C:$5n + 15$ (exp.) S:$n + 18$ (exp.)	C:$3n$ polynomial evaluations S: none	
Setup Comm. Complexity	$n(2.5n + 18.5)\lvert g\rvert$	C:$3n\lvert q\rvert$ S:$3\lvert q\rvert$	
Setup Comm. Round	4	C: 1 S: 1	
Retrieve Comp. Complexity	C:$14t + 24$ (exp.) S:$7t + 28$ (exp.)	C: 7 (exp.) S:$t + 10$ (exp.) G: 0 (exp.)	C: 7 (exp.) S:$3t + 10$ (exp.) G: 0 (exp.)
Retrieve Comm. Complexity	$(t + 1)(36.5 + 2.5n + 10.5(t + 1))\lvert g\rvert$	C:$5\lvert g\rvert$ S:$(4t + 5)\lvert g\rvert$ G:$(4t + 5)\lvert g\rvert$	C:$5\lvert g\rvert$ S:$(3t + 5)\lvert g\rvert + t\lvert q\rvert$ G:$(4t + 5)\lvert g\rvert$
Retrieve Comm. Rounds	10	C: 2 / S: 6 G: 4	C: 2 / S: 4 G: 4

From Table 7.1, it one can be seen that the retrieval protocol is significantly more efficient than the retrieval protocol of Camenisch et al., not only in communication rounds for the client but also in computation and communication complexities. In particular, the performance of the client in the retrieval protocol is independent of the number of servers and the threshold.

Efficiency of the TPASS Protocol Based on Zero-Knowledge Proof. The second protocol has the same initialization, client retrieval request, and client secret retrieval as the first protocol.

In the retrieval response, each server S_i participates in four communication rounds with other servers and the gateway, i.e., receiving msg_C from the gateway, broadcasting $\langle ID_C, B_i, C_i, D_i, \delta_i\rangle$ to other servers, receiving $\langle ID_C, B_j, C_j, D_j, \delta_j\rangle$ for all $j \neq i$ from other servers, and finally sending msg_i^* to the gateway.

The performance of the second protocol is also shown in Table 7.1, comparing with the performance of the Camenisch et al. (2014) protocol and the first protocol. One can notice that the retrieval computation complexity of the second protocol in a server is more than that in the first protocol, but is much less than that in the Camenisch et al. protocol. The retrieval communication complexity of the second protocol in a server is less than that in the first protocol because $\lvert q\rvert$ is less than $\lvert g\rvert$. In addition, the second protocol has reduced the communications among servers from two phases to one phase. The total communication overheads for each protocol is also much less than that in the Camenisch et al. protocol.

7.5 SECURITY ANALYSIS

7.5.1 Security Analysis of the TPASS Protocol Based on Two-Phase Commitment

Based on the security model described in the previous sections, we have the following theorem:

Theorem 1. Assuming that the decisional Diffie-Hellman (DDH) problem (Diffie and Hellman, 1976) is hard over $\{\mathbb{G}, q, g_1\}$, then the TPASS protocol P based on two-phase commitment illustrated in Figure 7.3, where the TPASS protocol is secure against the passive attack according to Definition 7.1.

Proof. In the security analysis, we consider the worst case where $t - 1$ servers have been corrupted. Without loss of generality, we assume that the first server S_1 is honest and the rest have been corrupted.

Given a passive adversary $\mathcal{A}$ attacking the protocol, we imagine a simulator S that runs the protocol for $\mathcal{A}$.

First of all, the simulator S initializes the system by generating public parameters $params = \{\mathbb{G}, q, g_1, g_2, H\}$. Next, Server $= \{S_1, S_2, \ldots, S_n\}$ and Client sets are determined. For each $C \in$ Client, a password pw_C and a secret s_C are chosen at random and then secret-shared with the n servers. In addition, the digest of the secret $H(s_C)$ is also secret-shared with the n servers.

The public parameters $params$ and the shares $\{ID_C, i, f_1(i), f_2(i), f_3(i)\}$ for $i = 2, 3, \ldots, t$ are provided to the adversary. When answering to any oracle query, the simulator S provides the adversary $\mathcal{A}$ with the internal state of the corrupted servers S_i ($i = 2, 3, \ldots, t$).

We refer to the real execution of the experiment, as described above, as P_0. A sequence of transformations is introduced to the experiment P_0 and bound the effect of each transformation to the adversary's advantage. We then bound the adversary's advantage in the final experiment. This immediately yields a bound on the adversary's advantage in the original experiment.

Experiment P_1. In this experiment, the simulator interacts with the adversary $\mathcal{A}$ as in experiment P_0 except that the adversary's queries to *Execute* oracles are handled differently: in any *Execute*$(C, i, \mathbb{S})$, where the adversary $\mathcal{A}$ has not queried Corrupt(C), the password pw_C in $msg_C = \langle ID_C, A \rangle$ where $A = g_1^r g_2^{\mathsf{pw}_C}$ is replaced with a random pw in $\mathbb{Z}_q^*$.

Because r in $A = g_1^r g_2^{\mathsf{pw}_C}$ is randomly chosen from $\mathbb{Z}_q^*$ by the simulator, the adversary cannot distinguish $g_1^r g_2^{\mathsf{pw}_C}$ with $g_1^r g_2^{\mathsf{pw}}$. Otherwise, we can break the semantic security of the ElGamal encryption, i.e., given an ElGamal encryption (g^r, my^r) where m is either $g_2^{pw_C}$ or g_2^{pw}, g is a generator of $\mathbb{G}$, and $y = g_1$, to determine if it is an encryption of $g_2^{pw_C}$. The semantic security of the ElGamal encryption is built on the DDH assumption. Therefore, we have:

Claim 1. If the DDH problem is hard over $\{\mathbb{G}, q, g_1\}$, $|\mathsf{Adv}_{\mathcal{PA}}^{P_0}(k) - \mathsf{Adv}_{\mathcal{PA}}^{P_1}(k)|$ is negligible.

Experiment P_2. In this experiment, the simulator interacts with the adversary $\mathcal{A}$ as in experiment P_1 except that: for any $Execute(C, i, \mathbb{S})$ oracle, where the adversary $\mathcal{A}$ has not queried $\mathsf{Corrupt}(C)$ and $\mathsf{Corrupt}(S_1)$, $a_1 f_1(1)$ in $\mathsf{msg}_1 = \langle ID_C, \delta_1, B_1, C_1, D_1 \rangle$ where $B_1 = g_1^{r_1} g_2^{a_1 f_1(1)}$ is replaced by a random number in $\mathbb{Z}_q^*$.

Because r_1 in $B_1 = g_1^{r_1} g_2^{a_1 f_1(1)}$ is randomly chosen from $\mathbb{Z}_q^*$ by the simulator, the adversary cannot distinguish $g_1^{r_1} g_2^{a_1 f_1(1)}$ from $g_1^{r_1} g_2^{\alpha}$, where α is a random number in $\mathbb{Z}_q^*$.

Claim 2. If the DDH problem is hard over $\{\mathbb{G}, q, g_1\}$, $|\mathsf{Adv}_{\mathcal{PA}}^{P_1}(k) - \mathsf{Adv}_{\mathcal{PA}}^{P_2}(k)|$ is negligible.

Experiment P_3. In this experiment, the simulator interacts with the adversary $\mathcal{A}$ as in experiment P_2 except that: for any $Execute(C, i, \mathbb{S})$ oracle, where the adversary $\mathcal{A}$ has not queried $\mathsf{Corrupt}(C)$ and $\mathsf{Corrupt}(S_1)$, E_1 in $\mathsf{msg}_1^* = \langle ID_C, C, D, E_1, F_1 \rangle$ is replaced with a random element in the group $\mathbb{G}$.

The difference between the current experiment and the previous one is bounded by the probability to solve the decisional Diffie-Hellman (DDH) problem over $\{\mathbb{G}, q, g_1\}$. More precisely, we have:

Claim 3. If the DDH problem over $\{\mathbb{G}, q, g_1\}$ is hard, $|\mathsf{Adv}_{\mathcal{PA}}^{P_2}(k) - \mathsf{Adv}_{\mathcal{PA}}^{P_3}(k)|$ is negligible.

If $|\mathsf{Adv}_{\mathcal{PA}}^{P_2}(k) - \mathsf{Adv}_{\mathcal{PA}}^{P_3}(k)|$ is non-negligible, we show that the simulator can use $\mathcal{A}$ as a subroutine to solve the DDH problem with non-negligible probability as follows.

Given a DDH problem (g_1^x, g_1^y, Z), where x, y are randomly chosen from $\mathbb{Z}_q^*$ and Z is either g_1^{xy} or a random element z from $\mathbb{G}$, the simulator replaces g_1^r in $A = g_1^r g_2^{pw_C}$ with g_1^x, $C_1 = g_1^{c_1}$ with g_1^y, and $(g_1^{c_1}, g_1^{c_1 r})$ in

$$E_1 = g_2^{a_1 f_2(1) h_1} g_1^{-r_1 \sum_{j=1}^{t} c_j} g_1^{c_1 \sum_{j=1}^{t} r_j} g_1^{c_1 r}$$

with g_1^y, Z, respectively, where r_j $(j = 1, 2, \ldots, t)$ and c_j $(j = 2, 3, \ldots, t)$ are randomly chosen by the simulator. When $Z = g^{xy}$, the experiment is the same as the experiment P_2. When Z is a random element z in $\mathbb{G}$, the experiment is the same as the experiment P_3. If the adversary can distinguish the experiments P_2 and P_3 with non-negligible probability, the simulator can solve the DDH problem with non-negligible probability.

Experiment P_4. In this experiment, the simulator interacts with the adversary $\mathcal{A}$ as in experiment P_3 except that: for any $Execute(C, i, \mathbb{S})$ oracle, where the adversary $\mathcal{A}$ has not queried Corrupt(C) and Corrupt(S_1), F_1 in $\mathsf{msg}_1^* = \langle ID_C, C, D, E_1, F_1 \rangle$ is replaced with a random element in the group $\mathbb{G}$.

Like the experiment P_3, we have

Claim 4. If the DDH problem is hard over $\{\mathbb{G}, q, g_1\}$, $|\mathsf{Adv}_{P\mathcal{A}}^{P_3}(k) - \mathsf{Adv}_{P\mathcal{A}}^{P_4}(k)|$ is negligible.

In experiment P_4, msg_C, msg_1, msg_1^* in *Execute* oracles have become independent of the password pw_C used by the client C and the secret s_C and $g_2^{H(s_C)}$ in the view of the adversary $\mathcal{A}$, if $\mathcal{A}$ does not require Corrupt(C) and Corrupt(S_1). In view of this, any off-line dictionary attack cannot succeed.

Experiment P_5. In this experiment, the simulator interacts with the adversary $\mathcal{A}$ as in experiment P_4 except that: for any $Execute(C, i, \mathbb{S})$ oracle, where the adversary $\mathcal{A}$ has not queried Corrupt(C) and Corrupt(S_1), the secret s_C of the client is replaced with a random element in the group $\mathbb{G}$.

Given a DDH problem (g_1^x, g_1^y, Z), where x, y are randomly chosen from $\mathbb{Z}_q^*$ and Z is either g_1^{xy} or a random element z from $\mathbb{G}$, the simulator replaces g_1^r in $A = g_1^r g_2^{\mathsf{pw}_C}$ with g_1^x, $C_1 = g_1^{c_1}$ with g_1^y, and $(g_1^r, g_1^{c_1 r})$ in

$$s_C = (E/C^r)^{h-1} = (E/(g_1^{r\sum_{i=2}^{t} c_i} g_1^{rc_1}))^{h-1}$$

with g_1^x, Z, respectively, where $h = H(ID_C, A, C, D)$, c_j ($j = 2, 3, \ldots, t$) are randomly chosen by the simulator. When $Z = g^{xy}$, the experiment is the same as the experiment P_4. When Z is a random element z in $\mathbb{G}$, the experiment is the same as the experiment P_5. If the adversary can distinguish the experiments P_4 and P_5 with non-negligible probability, the simulator can solve the DDH problem with non-negligible probability. Therefore, we have:

Claim 5. If the DDH problem is hard over $\{\mathbb{G}, q, g_1\}$, $|\mathsf{Adv}_{P\mathcal{A}}^{P_4}(k) - \mathsf{Adv}_{P\mathcal{A}}^{P_5}(k)|$ is negligible.

In experiment P_5, when the passive adversary $\mathcal{A}$ queries the $Test(C, i)$ oracle, the simulator S chooses a random bit b. When the adversary completes its execution and outputs a bit b', the simulator can tell whether the adversary succeeds by checking if (1) a $Test(C, i)$ query was made regarding some fresh client C and (2) $b' = b$.

The passive adversary's probability of correctly guessing the bit b used by the *Test* oracle is exactly 1/2 when the *Test* query is made to a fresh client instance C^i invoked by an *Execute*$(C, i, \mathbb{S})$ oracle. This is so because the secret s_C is chosen at random from $\mathbb{G}$, and hence there is no way to distinguish whether the *Test* oracle outputs a random secret or the "actual" secret (which is a random element anyway). Therefore, $\mathsf{Adv}_{P\mathcal{A}}^{P_5}(k) = 2 \cdot 1/2 - 1 = 0$. Based on Claims 1 to 5, we know $|\mathsf{Adv}_{P\mathcal{A}}^{P_0}(k) - \mathsf{Adv}_{P\mathcal{A}}^{P_5}(k)|$ is negligible. According to Definition 1, the protocol is secure against the passive attack and the theorem is proved. $\triangle$

Theorem 2.   Assuming that the DDH problem is hard over $\{\mathbb{G}, q, g_1\}$ and H is a collision-resistant hash function, then the TPASS protocol P based on a two-phase commitment illustrated in Figure 7.2 is secure against the active attack according to Definition 7.2.

Proof. Like in the proof of Theorem 1, we consider the worst case where $t - 1$ servers have been corrupted. The worst case can be divided into two subcases: (i) in the protocol, there is one honest server that has not been compromised by the adversary; (ii) all servers in the protocol are dishonest and controlled by the adversary.

For the first subcase,   without loss of generality, we assume that the first server S_1 is honest and the rest have been corrupted. Because of the inspection of the honest server, the published public parameters $\mathbb{G}, g_1, g_2, q$ cannot be changed and no one knows the discrete logarithm of g_2 based on g_1. Otherwise, it turns to the second subcase. This proof concentrates on the instances invoked by Send oracles. We view the adversary's queries to its Send oracles as queries to four different oracles as follows:

- Send(C, i) represents a request, for instance C^i of client C, to initiate the protocol. The output of this query is $\mathsf{msg}_C = \langle ID_C, A \rangle$.
- Send$(S_1, j, C, \mathsf{msg}_C)$ represents sending message msg_C to instance S_1^j of the server S_1, supposedly from the client C. The input of this query is $\mathsf{msg}_C = \langle ID_C, A \rangle$ and the output of this query is $\mathsf{msg}_1 = \langle ID_C, \delta_1, B_1, C_1, D_1 \rangle$.
- Send$(S_1, j, S_2, S_3, \ldots, S_t, M)$ represents sending message M to instance S_1^j of the server S_1, supposedly from the servers $S_2, S_3, \ldots, S_t$. The input of this query is $M = \mathsf{msg}_2 \parallel \mathsf{msg}_3 \parallel \ldots \parallel \mathsf{msg}_t$ and the output of this query is $\mathsf{msg}_1^* = \langle ID_C, C, D, E_1, F_1 \rangle$ or $\bot$.
- Send(C, i, msg_S) represents sending the message msg_S to instance C^i of the client C. The input of this query is $\mathsf{msg}_S = \langle ID_C, C, D, E, F \rangle$ and the output of this query is either $acc_C^i = TRUE$ or $\bot$.

Some terminologies will be used throughout the proof. A given message is called oracle-generated if it was output by the simulator in response to some oracle query. The message is said to be *adversarially generated* otherwise. An adversarially generated message must not be the same as any oracle-generated message.

An active adversary $\mathcal{A}$ is said to succeed if it makes an query Send(C, i, msg_S) to a fresh client instance C^i with **an adversarially generated message** msg_S, resulting in $acc_C^i = TRUE$. We denote this event by $\mathsf{Succ}_\mathcal{A}$.

Experiment P_6. In this experiment, the simulator interacts with the adversary as P_0 except that the adversary does not succeed, and the experiment is aborted, if any of the following occurs:

1. At any point during the experiment, an oracle-generated message (e.g., msg_C, msg_1, or msg_1^*) is repeated.
2. At any point during the experiment, a collision occurs in the hash function H (regardless of whether this is due to a direct action of the adversary or whether this occurs during the course of the simulator's response to an oracle query).

It is immediate that event 1 occurs with only negligible probability, while event 2 occurs with negligible probability assuming H as collision-resistant hash functions. Put everything together, we are able to see that:

Claim 6. If H is a collision-resistant hash function, $|\mathsf{Adv}_{\mathcal{A}\mathcal{A}}^{P_0}(k) - \mathsf{Adv}_{\mathcal{A}\mathcal{A}}^{P_6}(k)|$ is negligible.

Experiment P_7. In this experiment, the simulator interacts with the adversary $\mathcal{A}$ as in experiment P_6 except that (1) the adversary's queries to $\mathsf{Send}(C, i)$ oracles are handled differently: in any $\mathsf{Send}(C, i)$, where the adversary $\mathcal{A}$ has not queried $\mathsf{Corrupt}(C)$, the password pw_C in $\mathsf{msg}_C = \langle ID_C, A \rangle$ where $A = g_1^r g_2^{\mathsf{pw}_C}$ is replaced with a random number pw in $\mathbb{Z}_q^*$; (2) the adversary's queries to $\mathsf{Send}(S_1, j, C, \mathsf{msg}_C)$ oracles are handled differently: in any $\mathsf{Send}(S_1, j, C, \mathsf{msg}_C)$, where the adversary $\mathcal{A}$ has not queried $\mathsf{Corrupt}(C)$ and $\mathsf{Corrupt}(S_1)$, $a_1 f_1(1)$ in $\mathsf{msg}_1 = \langle ID_C, B_1, C_1, D_1 \rangle$ where $B_1 = g_1^{r_1} g_2^{a_1 f_1(1)}$ is replaced by a random number in $\mathbb{Z}_q^*$; (3) the adversary's queries to $\mathsf{Send}(S_1, j, S_2, \ldots, S_t, \mathsf{msg}_2 \| \ldots \| \mathsf{msg}_t)$ oracles are handled differently: in any $\mathsf{Send}(S_1, j, S_2, \ldots, S_t, \mathsf{msg}_2 \| \ldots \| \mathsf{msg}_t)$, where $\mathcal{A}$ has not queried $\mathsf{Corrupt}(C)$ and $\mathsf{Corrupt}(S_1)$, E_1 and F_1 in $\mathsf{msg}_1^* = \langle ID_C, C, D, E_1, F_1 \rangle$ are replaced with random elements in the group $\mathbb{G}$.

Like in experiments P_1 and P_2, the changes (1) and (2) will only bring a negligible change to the advantage of the active adversary.

For the change (3), if $\mathsf{msg}_C, \mathsf{msg}_2, \mathsf{msg}_3, \ldots, \mathsf{msg}_t$ are all oracle-generated, we can replace E_1 with a random element in $\mathbb{G}$ as in the experiment P_3.

If some of $\mathsf{msg}_C, \mathsf{msg}_2, \mathsf{msg}_3, \ldots, \mathsf{msg}_t$ are adversarially-generated, the adversary $\mathcal{A}$ cannot produce $A, (B_j, C_j, D_j)$ $(j = 2, 3, \ldots, t)$, such as $A \prod_{j=1}^{t} B_j$ excludes B_1 and $\delta_j = g_1^{H(ID_C, A, B_j, C_j, D_j)}$ for $j = 2, 3, \ldots, t$ still hold because A and the commitments δ_j $(j = 2, 3, \ldots, t)$ must be broadcast and received by the server S_1 at first and H is a collision-resistant hash function.

Because $B_1 = g_1^{r_1} g_2^{a_1 f_1(1)}$, we have

$$E_1 = g_2^{a_1 f_2(1) h_1} C^{-r_1} \left(A \prod_{j=1}^{t} B_j \right)^{c_i}$$

$$= g_2^{a_1 f_2(1) h_1} C^{-r_1} \left(A \prod_{j=2}^{t} B_j \right)^{c_i} g_1^{c_i r_1} (g_2^{c_i})^{a_1 f_1(1)}$$

Given a DDH problem (g_1^x, g_1^y, Z), where x, y are randomly chosen from $\mathbb{Z}_q^*$ and Z is either g_1^{xy} or a random element z from $\mathbb{G}$, the simulator replaces g_2 with g_1^x, $C_1 = g_1^{c_1}$ with g_1^y, and $(g_1^{c_1}, g_2^{c_1})$ in the above E_1 with g_1^y, Z, respectively, where r_1 is randomly chosen by the simulator. When $Z = g^{xy}$, the experiment is the same as the experiment P_6. When Z is a random element z in $\mathbb{G}$, the experiment is the same as the experiment P_7.

In the same way, we can make the change (4).

Because no one knows the discrete logarithm x of g_2 based on g_1, if the adversary can distinguish the experiments P_6 and P_7 with non-negligible probability, the simulator can solve the DDH problem with non-negligible probability. Therefore, we have:

Claim 7. If the DDH problem is hard over $\{\mathbb{G}, q, g_1\}$, $|\mathsf{Adv}_{\mathcal{A}\mathcal{A}}^{P_6}(k) - \mathsf{Adv}_{\mathcal{A}\mathcal{A}}^{P_7}(k)|$ is negligible.

In experiment P_7, msg_C, msg_1, msg_1^* in Send oracles have become independent of the password pw_C used by the client C and the secret s_C and $g_2^{H(s_C)}$ in the view of the adversary $\mathcal{A}$, if $\mathcal{A}$ does not require $\mathsf{Corrupt}(C)$ and $\mathsf{Corrupt}(\bar{S}_1)$. In view of this, any off-line dictionary attack cannot succeed.

To evaluate $Pr[\mathsf{Succ}_A]$, we consider three cases as follows.

Case 1. The adversary $\mathcal{A}$ forges $\mathsf{msg}_C' = \langle ID_C, A' \rangle$ where $A' = g_1^{r'} g_2^{\mathsf{pw}_C'}$ by choosing his own r' from $\mathbb{Z}_q^*$ and pw_C' from the dictionary D. In this case, if $Succ_A$ occurs, the adversary can conclude that the password used by the client is pw_C'. Therefore, the probability $Pr[Succ_A] = Q_1(k)/N$, where $Q_1(k)$ denotes the number of queries to $\mathsf{Send}(S_1, j, C, \mathsf{msg}_C)$ oracle.

Case 2. Given $\mathsf{msg}_C = \langle ID_C, A \rangle$, the adversary $\mathcal{A}$ forges $\mathsf{msg}_S = \langle ID_C, C', D', E', F' \rangle$ by choosing his own s', c', d' from $\mathbb{Z}_q^*$ and pw_C' from the dictionary D and computing $C' = g_1^{c'}, D' = g_1^{d'}, E' = g_2^{s'h'}(Ag_2^{\mathsf{pw}_C'})^{c'}, F' = g_2^{H(g_2^{s'})h'}(Ag_2^{\mathsf{pw}_C'})^{d'}$ where $h' = H(ID_C, A, C', D')$. When $\mathsf{pw}_C = \mathsf{pw}_C'$, we have $acc_C^i = TRUE$. Therefore, in this case, the probability $Pr[Succ_A] = Q_2(k)/N$, where $Q_2(k)$ denotes the number of queries to $\mathsf{Send}(C, i, \mathsf{msg}_S)$ oracle.

Case 3. Given $\mathsf{msg}_C = \langle ID_C, A \rangle$, the adversary $\mathcal{A}$ forwards msg_C to the server S_1 twice to get two responses $\langle ID_C, C_1, D_1, E_1, F_1 \rangle$ and $\langle ID_C, C_1', D_1', E_1', F_1' \rangle$. Then the adversary $\mathcal{A}$ sends to the client a forged message $\mathsf{msg}_S = \langle ID_C, C_1'/C_1, D_1'/D_1, g_2^{s^*h^*}E_1'/E_1, g_2^{H(g_2^{s^*})h^*}F_1'/F_1 \rangle$, where $E_1'/E_1 = g_1^{s(h'-h)}$, $(C_1'/C_1)^r, F_1'/F_1 = g_1^{H(g_1^s)(h'-h)}(D_1'/D_1)^r, h = H(ID_C, A, C, D), h' = H(ID_C, A, C', D'), h^* = H(ID_C, A, C_1'/C_1, D_1'/D_1)$, and s^* is chosen from $\mathbb{Z}_q^*$ by the adversary. The client accepts msg_S if and only if $h' = h$. Because H is a collision-resistant hash function, the probability $Pr[Succ_A]$ is negligible in this case.

In summary, in experiment P_7, $Pr[Succ_A] = Q(k)/N$, where $Q(k)$ denotes the number of on-line attacks. Based on Claims 6 and 7, we know $|\mathsf{Adv}_{P_A}^{P_0}(k) - \mathsf{Adv}_{P_A}^{P_7}(k)|$ is negligible. According to Definition 7.2, the protocol is secure against the active attack in the first subcase. $\triangle$

For the second subcase, all servers in the protocol are dishonest and controlled by the adversary, the active adversary can cheat the client with forged public parameters $\mathbb{G}, q, g_1, g_2$. However, the client can check if q is a large prime and $g_1^q = g_2^q = 1$ so that the discrete logarithm over $\mathbb{G}$ is hard although the adversary may know the discrete of g_2 based on g_1.

In this case, the active adversary can only query two Send oracles: Send(C, i) and Send(C, i, msg_S).

Experiment P_8. In this experiment, the simulator interacts with $\mathcal{A}$ as in experiment P_0 except that the adversary's queries to Send(C, i) oracles are handled differently: in any Send(C, i), where the adversary $\mathcal{A}$ has not queried Corrupt(C), the password pw_C in $\text{msg}_C = \langle ID_C, A \rangle$, where $A = g_1^r g_2^{\text{pw}_C}$ is replaced with a random number pw in $\mathbb{Z}_q^*$.

Like in experiment P_1, if the adversary can distinguish $g_1^r g_2^{\text{pw}_C}$ with $g_1^r g_2^{\text{pw}}$, we can break the semantic security of the ElGamal encryption. Therefore, we have:

Claim 8. If the DDH problem is hard over $\{\mathbb{G}, q, g_1\}$, $|\text{Adv}_{\mathcal{A}\mathcal{A}}^{P_0}(k) - \text{Adv}_{\mathcal{A}\mathcal{A}}^{P_8}(k)|$ is negligible.

In experiment P_8, the adversary can only perform the active attack as described in Case 2 of experiment P_7. In this case, the probability $Pr[Succ_A] = Q_2(k)/N$, where $Q_2(k)$ denotes the number of queries to Send(C, i, msg_S) oracle.

Based on Claim 8, we know $|\text{Adv}_{P\mathcal{A}}^{P_0}(k) - \text{Adv}_{P\mathcal{A}}^{P_8}(k)|$ is negligible. According to Definition 7.2, the protocol is secure against the active attack in the second subcase.

This completes the proof of the theorem. $\triangle$

7.5.2 Security Analysis of the TPASS Protocol Based on Zero-Knowledge Proof

Before analyzing the security of the second protocol, we introduce an assumption about noninteractive zero-knowledge proof of knowledge as follows.

Noninteractive Zero-Knowledge Proof of Knowledge Assumption. Considering a protocol where Prover, wishing to prove to Verifier that he knows x such that $y = g^x$, sends $(R = g^r, \alpha = H(g, y, R)x + r)$ (where r is randomly chosen by Prover) to Verifier, who accepts the proof if $g^\alpha = y^{H(g,r,R)}R$ and otherwise rejects the proof. We assume that in this protocol, Prover has to know x to generate the noninteractive zero-knowledge proof of knowledge (R, α) such that $g^\alpha = y^{H(g,r,R)}R$ and Verifier gains no knowledge of x after the proof.

Theorem 3. Assuming that the DDH problem is hard over $\{\mathbb{G}, q, g_1\}$ and the non-interactive zero-knowledge proof of knowledge assumption holds, the TPASS protocol P based on the zero-knowledge proof illustrated in Figure 7.3 is secure against the passive attack according to Definition 1.

Proof. The proof is the same as that of Theorem 1, except or Claims 3 and 4.

Claim 3'. If the DDH problem over $\{\mathbb{G}, q, g_1\}$ is hard and the noninteractive zero-knowledge proof of knowledge assumption holds, $|\mathsf{Adv}_{\mathcal{P_A}}^{P_2}(k) - \mathsf{Adv}_{\mathcal{P_A}}^{P_3}(k)|$ is negligible.

If $E_1 = g_2^{a_1 f_2(1) h_1} C^{-r_1} (A \prod_{j=1}^{t} B_j)^{c_i}$ contains $B_1 = g_1^{r_1} g_2^{a_1 f_1(1)}$, Claim 3' can be proved to be true in the same way as the proof of Theorem 1. If E_1 does not contain B_1, E_1 must contain $g_1^{r_1 c_1}$ because $C = g_1^{\sum c_j}$ and the zero-knowledge proof of knowledge assumption ensures the server S_j knowing c_j such that $C_j = g_1^{c_j}$. Given $C_1 = g_1^{c_1}$ and $g_1^{r_1} = B_1 g_2^{\sum_{j \neq 1} a_j f_1(j)} / g_2^{\mathsf{pw}}$ (where pw is chosen for the offline dictionary attack), the $t - 1$ servers $S_2, \ldots, S_t$ cannot distinguish $g_1^{r_1 c_1}$ with a random group element due to the DDH assumption. Note that the zero-knowledge proof of knowledge assumption ensures the server S_j ($j \neq 1$) having no knowledge of c_1. Therefore, Claim 3' is true.

Claim 4'. If the DDH problem over $\{\mathbb{G}, q, g_1\}$ is hard and the non-interactive zero-knowledge proof of knowledge assumption holds, $|\mathsf{Adv}_{\mathcal{P_A}}^{P_3}(k) - \mathsf{Adv}_{\mathcal{P_A}}^{P_4}(k)|$ is negligible.

The proof can be obtained from the proof of Claim 3' by replacing E_1, C, C_j, c_j with F_1, D, D_j, d_j. Please note that (C_j, D_j, δ_j) is also a noninteractive zero-knowledge proof of knowledge of d_j. $\triangle$

Theorem 4. Assuming that the DDH problem is hard over $\{\mathbb{G}, q, g_1\}$, the noninteractive zero-knowledge proof of knowledge assumption holds, and H is a collision-resistant hash function, then the TPASS protocol P based on the zero-knowledge proof illustrated in Figure 7.3 is secure against the active attack according to Definition 7.2.

Theorem 4 can be proved by combining the proofs of Theorems 1 to 3.

7.6 EXPERIMENTS

Experiments have been carried out to validate the performance of the two protocols. The experiments are performed with the following hardware specifications: CPU: 2.2 GHz Intel Core i7, Memory: 16 GB 1600 MHz DDR3.

The two protocols were implemeted with JRE System Library [JavaSE - 1.7] in the settings where there are 3, 5, 10, 15, 20 servers, respectively. In our experiments, the size of a group element (i.e, $|g|$) has 1024 bits and the order of the group (i.e, $|q|$) has 160 bits. One modular exponentiation for the 160-bit exponent takes approximately 0.0028 seconds and one modular multiplication takes approximate 0.000005 seconds.

7.6.1 Performance of Initialization

For $n = 3,5,10,15,20$, the performance of the two protocols per client in the initialization, compared with the Camenisch et al. (2014) protocol, is illustrated Figures 7.4 and 7.5. Because the two protocols have the same initialization, their performances

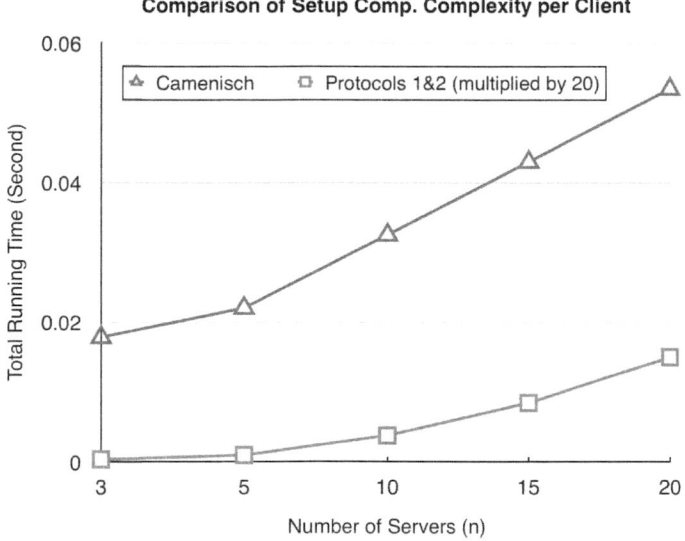

Figure 7.4 Comparison of time spent (in seconds) for setting up.

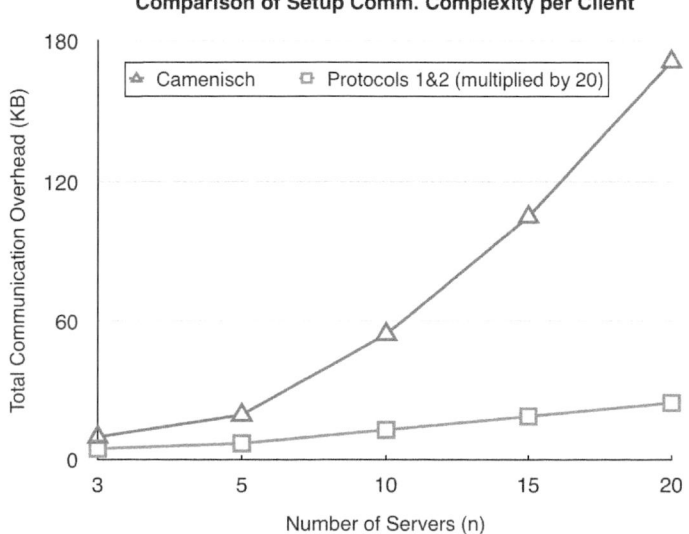

Figure 7.5 Comparison of communication size (in KB) for setting up.

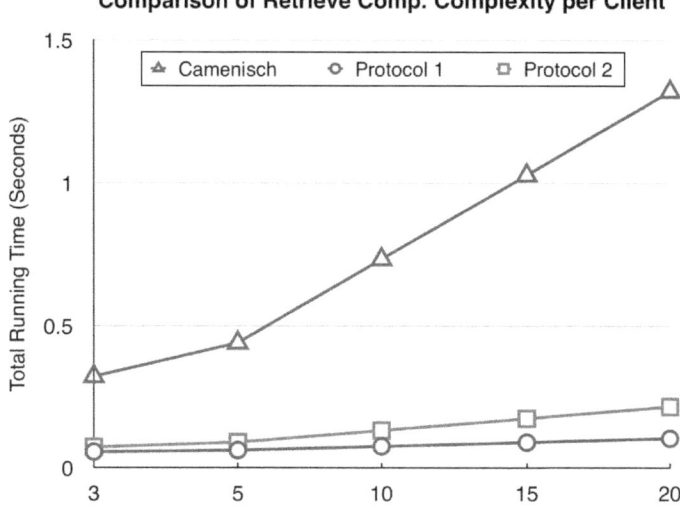

Figure 7.6 Comparison of time spent for retrieving.

are the same. In order to make the performance of the protocols visible in the comparison, the computational and communication complexities of the protocols for a client have been enlarged by 20 times.

As can be seen in Figures 7.4 and 7.5, the two described protocols are significantly more efficient than Camenisch et al.'s protocol and the difference increases with the increase in the number of servers (n).

7.6.2 Performance of Retrieve

For $t = 3,5,10,15,20$, the performance of the two protocols per client in the retrieval, compared to the Camenisch et al. (2014) protocol, is illustrated in Figures 7.6 and 7.7.

One can notice in Figure 7.6 that the total running time of the two protocols per client is less than the Camenisch et al.'s protocol. Around 95% was saved in the first protocol and 85% in the second protocol. Although the difference is just a couple of seconds for a client, it will become significantly large when the servers provide services to a large number of clients concurrently. If we ignore the communication time, the verification of the second protocol is a little bit slower that the first protocol. However, if the communication time cannot be ignored, the second protocol may be more efficient than the first protocol because it reduces the communications of the servers from two phases to one phase.

The communication overhead of Camenisch et al.'s protocol depends on the total number n of servers. In Figure 7.7, we assume $n = t + 1$. In addition, we assume that

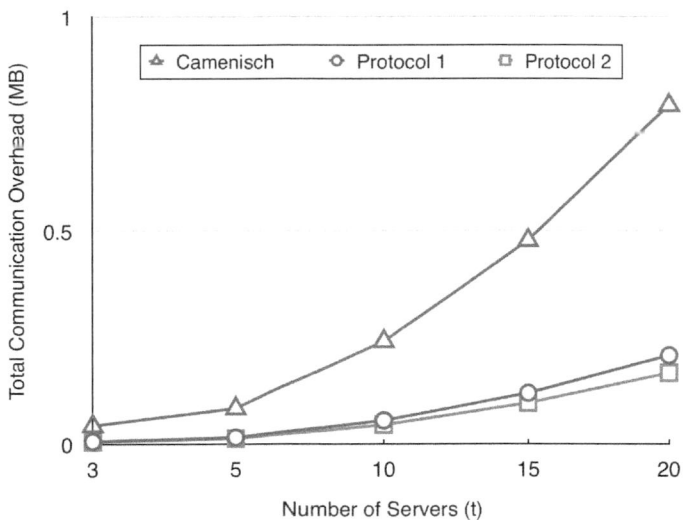

Figure 7.7 Comparison of communication size for retrieving.

the broadcast channel in the two protocols is implemented by point to point communication. In this case, the total communication overheads of the two protocols are $(4t^2 + 5t)|g|$ bits and $(3t^2 + 5t)|g| + t^2|q|$ bits, respectively.

Once can notice also in Figure 7.7 that the two protocols have almost the same communication overheads, which is significantly less than Camenisch et al.'s protocol. Up to 65% is saved up in the first protocol and 75% in the second protocol.

In addition, for $t = 10$, the performance of the two protocols per server in the retrieval, compared with Camenisch et al.'s protocol, is illustrated in Figures 7.8 and 7.9, which also show that the described protocols are more efficient than Camenisch et al.'s protocol.

7.7 CONCLUSION

This chapter presented two efficient t-out-of-n TPASS protocols for any $n > t$ that protects the password of the client when he/she tries to retrieve the secret from all corrupt servers as well as prevent the adversary from planting a different secret into the user's mind than the secret that was stored earlier. The described protocols are significantly more efficient than existing TPASS protocols. Furthermore, a rigorous proof of security for the protocols in the standard model is provided in this chapter, followed by a series of experimental results. Future work could study how efficiently the corrupted servers could be detected and the protocol in lightweight mobile devices could be implemented to support cloud-based services.

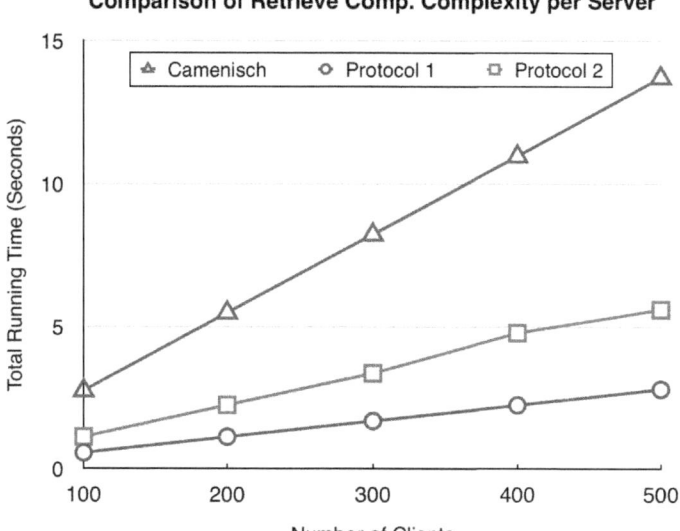

Figure 7.8 Comparison of average time spent by a server in retrieving.

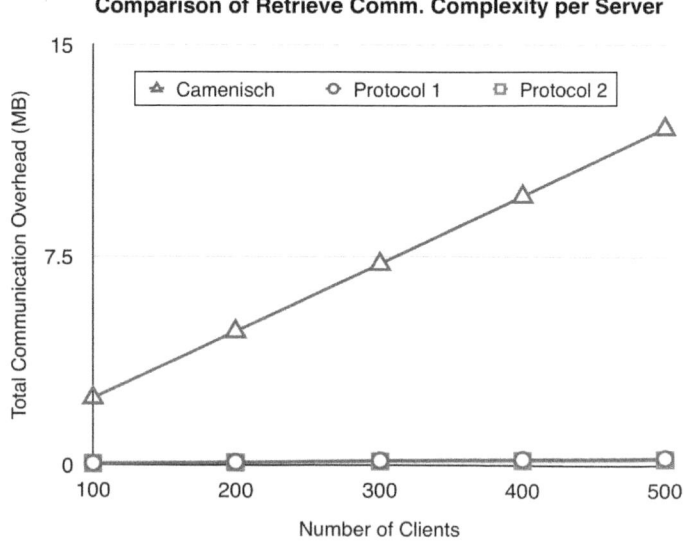

Figure 7.9 Comparison of average communication size for a server in retrieving.

CHAPTER 8

Conclusion

Intrusion Detection Systems (IDSs) have become a promising security method for anomaly detection in traditional IT systems. The increased importance of this method has opened up an interesting research area in security, and its use is not confined to traditional IT systems, but has been adapted to detect unexpected behaviors in SCADA systems. However, the different nature and characteristics of SCADA systems have motivated security researchers to develop SCADA-specific IDSs. *Information sources* (e.g., network-based, application-based) through which anomalous behaviors can be detected and *modeling methods* that are used to build the detection models using the available information source are the two salient parts that must be taken into account in the development of IDSs.

A number of IDSs have been specifically designed to meet the requirements of SCADA systems. However, they vary with respect to the modeling methods used to build the detection model, and in addition to the selected information source. Significant recent research has revealed that a SCADA data-driven IDS has the potential to revolutionize the secure monitoring of our daily infrastructures and industrial processes. The building of detection models using SCADA data can be performed using two frequently used strategies: manual definition or machine learning. The former is time-consuming and prohibitively expensive as it requires experts, while the latter requires the machine to automatically build a detection model based on a set of training data. However, the detection accuracy of any created detection model relies totally on the goodness of that model. By "goodness" we refer to the ability of the detection model to distinguish between normal and abnormal behavior. This can be achieved by training the anomaly detection methods with a labeled training data set, where each observation is denoted as either normal or anomalous. This type of learning is called **supervised mode**. However, labeled data should be accurately labeled and comprehensively represent all types of behaviors, which makes it prohibitively expensive because this requires domain experts to label hundreds of thousands of data observations. In addition, it is challenging to obtain abnormal observations that comprehensively represent anomalous behaviors.

To address the previously discussed issues in the supervised mode, the **semi-supervised** mode is proposed where the anomaly detection method is trained

SCADA Security: Machine Learning Concepts for Intrusion Detection and Prevention,
First Edition. Abdulmohsen Almalawi, Zahir Tari, Adil Fahad and Xun Yi.

with just a one-class problem of either normal or abnormal data. Obtaining a normal training data set can be done by running a target system under normal conditions and the collected data is assumed to be normal. However, it is difficult to obtain purely normal data that comprehensively represent normal behaviors. On the another hand, it is challenging to obtain a training data set that covers all possible abnormal behavior that could occur in the future. Therefore, the **unsupervised mode**, where the labeled data is not available, has the potential to address the aforementioned issues of the previous modes, namely supervised and semisupervised modes. However, low accuracy and efficiency of the unsupervised mode have made it a poor alternative. In this thesis, we were motivated in particular to develop a robust, unsupervised SCADA-specific IDS that is based on SCADA data. A set of solutions has been proposed to achieve this aim.

The proposed unsupervised SCADA data-driven IDS and framework for the SCADA security testbed can facilitate the nearly optimal design of SCADA systems in which the intrusion behaviour that deviates from an expected pattern (the predened normal behavior) can be easily monitored. For instance, a predetermined threshold for a single parameter is the most commonly used approach to monitor a system's behaviors, where any value exceeding this threshold is considered as anomalous. However, the value of a single process parameter may not be abnormal, but, in combination with other process parameters, may produce abnormal behaviors, which very rarely occurs. Therefore, the described unsupervised SCADA data-driven IDS can be used to find out the nearly optimal combination of parameters that can nearly define the boundaries of hidden normal behaviors of a targeted system. Moreover, the proposed framework for the SCADA security testbed can be used to design the proposed SCADA system, in addition to simulating security measures and testing attack scenarios prior to implementing the proposed system in the real world.

8.1   SUMMARY

The following specific research problems have been addressed in this book, namely:

A. How to develop a SCADA testbed that is a realistic alternative for real SCADA systems so that it can be used for efficient SCADA security evaluation and testing purposes.

B. How to efficiently find the k-NN in large and high-dimensional data.

C. How to learn, from unlabeled SCADA data, clustering-based proximity rules for SCADA anomaly detection.

D. How to find a global and efficient anomaly threshold for unsupervised detection.

The first problem focused on the development of a framework for a SCADA security testbed that is intended to be an evaluation and testing environment for SCADA security in general, and in particular for the described unsupervised IDS. The following three research questions focus on the development of methods that are used together to build a robust unsupervised SCADA data-driven IDS. In the following, the overall thesis contributions are summarized.

A Framework for SCADA Security Testbed (SCADAVT)

The establishment of a SCADA research laboratory is very important in identifying common attacks and evaluating security solutions tailored to SCADA systems. However, it is impractical to evaluate SCADA security solutions on actual live systems. Thus, a framework for a SCADA security testbed based on virtualization technology has been proposed. In this framework, the two main parts, namely *main SCADA system components* and *controlled environment*, have been modeled in order to realistically mimic an actual SCADA system. In the former part, all the essential SCADA components (e.g., MUT, PLC, ModBus protocol) have been built on top of the CORE emulator that has been used as a communication infrastructure for SCADA components. The integration of the essential SCADA components into a CORE emulator introduces a friendly interface to create a SCADA system that is configurable for any application to control and supervise. In the latter part, the server, which acts as a surrogate for water distribution systems, has been introduced to simulate water movement and quality behavior within pressurized pipe networks using the well-known modeling dynamic link library (DLL) of EPANET. In addition, the server can simulate any topology of water network systems and can be manipulated by a custom TCP-based protocol. A case study is presented to demonstrate how the testbed can easily be used and configured to monitor and control any automatised processes. Moveover, DDoS and integrity attacks have been described to illustrate how malicious attacks can be launched on supervised processes.

An Efficient Search for *k*-NN in Large and High-Dimensional Data

We undertook this study to develop a SCADA data-driven IDS that operates in unsupervised mode where labels are not available. It is a challenging process to build detection models from unlabeled data. This is because an unlabeled data set cannot be guaranteed to represent one behavior as either normal or abnormal. The separation between normal and abnormal data is the salient part in the training phase of the proposed system. The k-Nearest Neighbour-based(k-NN) approach is found in the literature to be one of the most interesting and best approaches for mining anomalies. However, a large computational cost is incurred when finding the k-NN for any query object because all objects in a data set must be checked. Therefore, an efficient k-nearest-neighbour-based approach was presented, called kNNVWC, which is based on a novel various-widths clustering algorithm and triangle inequality. This approach makes possible the partitioning of a data set using various widths whereby each width suits a particular distribution. The novel various-widths clustering method for partitioning a data set has introduced two features: (i) the balance between the number of produced clusters and their respective sizes has maximized the efficiency of using triangle inequality to prune unlikely clusters and (ii) clustering with a global width first produces relatively large clusters, and each cluster is independently partitioned without considering other clusters, thereby reducing the clustering time.

In the experiments, 12 data sets, which vary in domains, size, and dimensionality, have been used. The experimental results have shown that kNNVWC outperforms four algorithms in both construction and search times. Moveover, kNNVWC has shown promising results with high-dimensional data in searching k-NN for a query object.

Clustering-Based Proximity Rules for SCADA Anomaly Detection

The separation between normal and abnormal behavior using the efficient proposed kNNVWC approach needs a further process in order to build an efficient SCADA data-driven detection model. This is because it is impractical to use all the training data in the anomaly detection phase, as a large memory capacity is needed to store all scored observations, and it is computationally infeasible to calculate the similarity between these observations and each current new observation. Consequently, a novel SCADA Data-Driven Anomaly Detection approach, called SDAD, was introduced based on a clustering-based method to extract proximity-based detection rules. This has been performed by initially separating the inconsistent observations from consistent ones, and then the proximity detection rules for each behavior, whether consistent or inconsistent, are automatically extracted from the observations that belong to that behaviour. The extracted rules are used to monitor and warn of inconsistent observations that are produced by the target SCADA data points for a given SCADA system. The extracted proximity-based detection rules have demonstrated promising results for detecting anomalies in SCADA data. Moreover, we also tested the reliability of this approach with the existing clustering-based intrusion detection algorithm.

Towards Global Anomaly Threshold to Unsupervised Detection

An unsupervised intrusion detection method is an appropriate method for detecting anomalies in unlabeled data. This is because the labeling of the huge amount of data produced by SCADA systems is a costly and time-consuming process. However, unsupervised learning algorithms are based on some assumptions by which anomalies are detected, where these assumptions are controlled by several parameter choices. Thus, Global Anomaly Threshold to Unsupervised Detection, called GATUD, has been proposed. This approach is proposed as an add-on component to improve the accuracy of unsupervised intrusion detection methods. This has been performed by initially learning two labeled small data sets from the unlabeled data, where each data set represents either normal or abnormal behavior. Then, a set of supervised classifiers are trained with these two data sets to build an ensemble-based decision-making model that can be integrated into both unsupervised anomaly scoring and clustering-based intrusion detection approaches. In the former, GATUD is used to find a global and efficient anomaly threshold, while in the latter it is used to efficiently label produced clusters as either normal or abnormal. Experiments have shown that the integration of GATUD into the proposed SDAD approach in Chapter 3 has demonstrated significant results. Moreover, GATUD demonstrated significant and promising results when it was integrated into a clustering-based intrusion detection approach as a labeling method for the produced clusters.

8.2 FUTURE WORK

This book proposed a set of innovative solutions/methods that have been used together to make a robust unsupervised SCADA data-driven IDS. In addition, a framework for a realistic SCADA testbed has been introduced. Nevertheless, important future work still needs to be done to refine each method.

- In Chapter 3, a framework for a SCADA security testbed based on virtualization technology was described. In this framework, both *main SCADA system components* and a *controlled environment* have been considered so that a user can fully simulate a full SCADA system. However, this framework is still limited to Modbus/TCP-based components. That is, all the simulated SCADA components, such as PLC, MTU server, HMI Server, etc., are Modbus/TCP-based. Therefore, it is valuable to integrate some notable SCADA protocols such as Zigbee and DNP3. A server that is used as a surrogate for water distribution systems has been introduced in this framework, although it would be ideal to integrate other controlled environments such as Power-World and wind turbine simulators. This integration can be easily performed by extending the generic gateway provided in this framework.

- In Chapter 4, an approach to efficiently search for k-NN in large and high-dimensional data has been proposed. This approach has demonstrated significant results compared to well-known algorithms and has made it possible to efficiently give an anomaly score for each observation based on its neighborhood density, even with large and high-dimensional data. However, the efficiency of this approach depends on the efficiency of the novel various-widths clustering algorithm, which partitions a data set into a number of clusters using various widths. The number and size of the produced clusters are controlled by the partitioning and merging processes in this algorithm. The creation of a small number of clusters can result in large clusters that incur a high computational cost when finding k-NN for an observation. On the other hand, a large number of produced clusters can result in a large number of distance computations between an observation and the centroids of the produced clusters in order to prune unlikely clusters. Thus, the optimization of the optimal balance of the number of produced clusters and their respective sizes is significantly important and should be considered in future work.

- Chapter 5 described a new approach that extracts, from unlabeled SCADA data, proximity detection rules based on the clustering method for SCADA anomaly detection. These proximity detection rules can be extracted from large and high-dimensional SCADA data, where the k-nearest neighbor approach, that was described in Chapter 3, is used to efficiently give an anomaly score for each observation based on its neighborhood density. The normal proximity detection rules are then extracted from observations that are strongly expected to represent the normal behavior. From the abnormal observations, abnormal proximity detection rules were extracted. Although the results of the extracted proximity detection rules are promising, there is a need to dynamically update these rules, as the normal behaviors of a given system may evolve over time.

- Although the SDAD approach proposed in Chapter 5 has the ability to extract, from unlabeled data, proximity detection rules based on clustering, it suffers from low detection accuracy. This is because the detection of anomalies in this mode (unsupervised) is based on assumptions to find the near-optimal anomaly threshold. Therefore, the best detection accuracy is achieved when a near-optimal choice of the anomaly threshold is met. Chapter 6 described an approach that can be used as an add-on component to find a global and efficient anomaly threshold. Two major steps are involved: learning of a most-representative data set from unlabeled data, which represents both normal and abnormal behaviors, and building an ensemble-based decision-making model. Although this approach has shown promising results, there is still scope for future work: (i) minimizing the size of the most-representative data set as much as possible without losing its representative characteristics for the learning data and (ii) evaluating this approach with further intrusion-detection methods for various application domains.

REFERENCES

M. Abdalla, M. Cornejo, A. Nitulescu, and D. Pointcheval. Robust password-protected secret sharing. In *Proceedings of European Symposium on Research in Computer Security (ESORICS)*, pages 61–79, 2016.

J. Ahrenholz. Comparison of core network emulation platforms. In *Proceedings of IEEE MILCOM Conference*, pages 166–171, 2010.

J. Ahrenholz. CORE documentation, March 2014. URL: http://downloads.pf.itd.nrl.navy.mil/docs/core/core_manual.pdf.

Jeff Ahrenholz, Tom Goff, and Brian Adamson. Integration of the core and emane network emulators. In *Proc. of IEEE Military Communications Conference*, pages 1870–1875, 2011.

Cristina Alcaraz and Javier Lopez. Diagnosis mechanism for accurate monitoring in critical infrastructure protection. *Elsevier Journal on Computer Standards and Interfaces*, 36 (3): 501–512, 2014a.

Cristina Alcaraz and Javier Lopez. Wasam: A dynamic wide-area situational awareness model for critical domains in smart grids. *Elsevier Journal on Future Generation Computer Systems*, 30: 146–154, 2014.

Abdulmohsen Almalawi, Adil Fahad, Zahir Tari, Abdullah Alamri, Rayed AlGhamdi, and Albert Y Zomaya. An efficient data-driven clustering technique to detect attacks in SCADA systems. *IEEE Transactions on Information Forensics and Security*, 11 (5): 893–906, 2015.

American Gas Association (AGA). Cryptographic protection of SCADA communications part 1: Background, policies and test plan. Technical report, Technical Report AGA Report, 2006.

Fabrizio Angiulli and Fabio Fassetti. Dolphin: An efficient algorithm for mining distance-based outliers in very large datasets. *ACM Transactions on Knowledge Discovery from Data (TKDD)*, 3 (1): 4, 2009.

Fabrizio Angiulli and Clara Pizzuti. Fast outlier detection in high dimensional spaces. *Springer Journal on Principles of Data Mining and Knowledge Discovery*, pages 43–78, 2002.

Fabrizio Angiulli, Stefano Basta, and Clara Pizzuti. Distance-based detection and prediction of outliers. *IEEE Transactions on Knowledge and Data Engineering (TKDE)*, 18 (2): 145–160, 2006.

Mikhael Ankerst, Markus M. Breunig, Hans-Peter Kriegel, and Jörg Sander. Optics: Ordering points to identify the clustering structure. *ACM SIGMOD Rec.*, 28 (2): 49–60, June 1999.

Andreas Arning, Rakesh Agrawal, and Prabhakar Raghavan. A linear method for deviation detection in large databases. pages 164–169. *Proceedings of International Conference on Knowledge Discovery and Data Mining (KDD)*, August 1996.

SCADA Security: Machine Learning Concepts for Intrusion Detection and Prevention,
First Edition. Abdulmohsen Almalawi, Zahir Tari, Adil Fahad and Xun Yi.
© 2021 John Wiley & Sons, Inc. Published 2021 by John Wiley & Sons, Inc.

S. J Jarecki, A. Kiayias, H. Krawczyk, and J. Xu. Toppss: Cost-minimal password-protected secret sharing based on threshold oprf. In *Proceedings of Applied Cryptography and Network Security (ACNS)*, pages 39–58, 2017.

A. Bagherzandi, S. Jarecki, N. Saxena, and Y. Lu. Password-protected secret sharing. In *Proceedings of the ACM Conference on Computer and Communications Security (CSS)*, pages 433–444, 2011.

Stephen D. Bay and Mark Schwabacher. Mining distance-based outliers in near linear time with randomization and a simple pruning rule. In *Proceedings of the 9th ACM SIGKDD International Conference on Knowledge Discovery and Data Mining*, pages 29–38, 2003.

M. Bellare, D. Pointcheval, and P. Rogaway. Authenticated key exchange secure against dictionary attacks. In *Proceedings of Eurocrypt*, pages 139–155, 2000.

Boldizsár Bencsáth, Gábor Pék, Levente Buttyán, and Márk Félegyházi. Duqu: Analysis, detection and lessons learned. In *ACM European Workshop on System Security (EuroSec)*, volume 2012, 2012.

Alina Beygelzimer, Sham Kakade, and John Langford. Cover trees for nearest neighbor. In *Proceedings of the 23rd ACM International Conference on Machine Learning*, pages 97–104, 2006. ISBN 1-59593-383-2.

Jim Bird and Frank Kim. SANS survey on application security programs and practices, 2012. URL: http://www.sans.org/reading_room/analysts_program/sans_survey_appsec.pdf.

Stuart A. Boyer. *SCADA: Supervisory Control And Data Acquisition*. International Society of Automation, USA, 4th edition, 2009.

J. Brainard, A. Juels, B. Kaliski, and M. Szydlo. Nightingale: A new two-server approach for authentication with short secrets. In *Proceedings of the 12th USENIX Security Symposium*, pages 201–213, 2003.

M. M. Breunig, H. Kriegel, R. T. Ng, and J. Sander. Lof: Identifying density-based local outliers. In *Proceedings of the ACM SIGMOD International Conference on Management of Data*, pages 93–104, 2000.

Markus M. Breunig, Hans-Peter Kriegel, Raymond T. Ng, and Jörg Sander. Optics-of: Identifying local outliers. In *Proceedings of the 3rd Springer European Conference on Principles of Data Mining and Knowledge Discovery*, pages 262–270, 1999.

Christopher Bronk and Eneken Tikk-Ringas. The cyber attack on Saudi Aramco. *Survival*, 55 (2): 81–96, 2013.

Eric Byres, John Karsch, and Joel Carter. NISCC good practice guide on firewall deployment for SCADA and process control networks. *National Infrastructure Security Co-Ordination Centre*, 2005.

J. Camenisch, A. Lysyanskaya, and G. Neven. Practical yet universally composable two-server password-authenticated secret sharing. In *Proceedings of the ACM Conference on Computer and Communications Security (CSS)*, pages 525–536, 2012.

J. Camenisch, A. Lehmann, A. Lysyanskaya, and G. Neven. Memento: How to reconstruct your secrets from a single password in a hostile environment. In *Proceedings of the Crypto Conference*, pages 256–275, 2014.

A. Carcano, A. Coletta, M. Guglielmi, M. Masera, I. Nai Fovino, and A. Trombetta. A multidimensional critical state analysis for detecting intrusions in SCADA systems. *IEEE Transactions Industrial Informatics (TII)*, 7 (2): 179–186, May 2011.

Varun Chandola, Arindam Banerjee, and Vipin Kumar. Anomaly detection: A survey. *ACM Computing Surveys (CSUR)*, 41 (3): 15, 2009.

C.Y. Chen, Chin-Chen Chang, and Richard Chia-Tung Lee. A near pattern-matching scheme based upon principal component analysis. *Elsevier Pattern Recognition Letters*, 16 (4): 339–345, 1995.

Steven Cheung, Bruno Dutertre, Martin Fong, Ulf Lindqvist, Keith Skinner, and Alfonso Valdes. Using model-based intrusion detection for SCADA networks. In *Proceedings of the SCADA Security Scientific Symposium*, pages 127–134, 2007.

Henrik Christiansson and Eric Luiijf. Creating a European SCADA security testbed. In *Proceedings IFIP International Federation for Information Processing*, volume 253, pages 237–247, 2010.

Brent Chun, David Culler, Timothy Roscoe, Andy Bavier, Larry Peterson, Mike Wawrzoniak, and Mic Bowman. Planetlab: An overlay testbed for broad-coverage services. *ACM SIG-COMM Computer Communication Review*, pages 3–12, 2003.

Thomas Cover and Peter Hart. Nearest neighbor pattern classification. *IEEE Transactions on Information Theory*, 13 (1): 21–27, 1967.

André Cunha, Anis Koubaa, Ricardo Severino, and Mário Alves. Open-zb: An open-source implementation of the ieee 802.15. 4/zigbee protocol stack on tinyos. In *IEEE Internatonal Conference on Mobile Adhoc and Sensor Systems*, pages 1–12, 2007.

C. M. Davis, J. E. Tate, H. Okhravi, C. Grier, T. J. Overbye, and D. Nicol. SCADA cyber security testbed development. In *Proceedings of the 8th IEEE North American Power Symposium (NAPS)*, pages 483–488, 2006.

Dorothy E. Denning. An intrusion-detection model. *IEEE Transactions on Software Engineering (TSE)*, SE-13 (2): 222–232, Feb. 1987.

QualNet Developers. Scalable network technologies: Qualnet developer, July 2012. URL: http://www.scalable-networks.com/products/developer.php.

Thomas G. Dietterich. Ensemble methods in machine learning. *Springer Journal on Multiple Classifier Systems*, pages 1–15, 2000.

W. Diffic and M. Hellman. New directions in cryptography. *IEEE Transactions on Information Theory*, 32 (2): 644–654, 1976.

Digitalbond. IDS-signatures/modbus-tcp, July 2013. URL: http://www.digitalbond.com/index .php/research/ids-signatures/modbus-tcp-ids-signatures/.

T. ElGamal. A public-key cryptosystem and a signature scheme based on discrete logarithms. *IEEE Transactions on Information Theory*, 31 (4): 469–472, 1985.

Eleazar Eskin, Andrew Arnold, Michael Prerau, Leonid Portnoy, and Sal Stolfo. A geometric framework for unsupervised anomaly detection. In *Applications of Data Mining in Computer Security*, pages 77–101. Springer, 2002a.

Eleazar Eskin, Andrew Arnold, Michael Prerau, Leonid Portnoy, and Salvatore Stolfo. *A Geometric Framework for Unsupervised Anomaly Detection: Detecting Intrusions in Unlabeled Data*. Kluwer, 2002b.

N. Falliere, L. O. Murchu, and E. Chien. W32.stuxnet dossier. Technical report. Technical Report 1.4, Symantec Tech, 2011.

Eduardo B. Fernandez, Jie Wu, M. M. Larrondo-Petrie, and Yifeng Shao. On building secure SCADA systems using security patterns. In *Proceedings of the 5th ACM Annual Workshop on Cyber Security and Information Intelligence Research: Cyber Security and Information Intelligence Challenges and Strategies*, page 17, 2009.

W. Ford and B. S. Kaliski. Server-assisted generation of a strong secret from a password. In *Proceedings of the 5th IEEE Intl. Workshop on Enterprise Security*, 2000.

Igor Nai Fovino, Andrea Carcano, Thibault De Lacheze Murel, Alberto Trombetta, and Marcelo Masera. Modbus/dnp3 state-based intrusion detection system. In *Proceedings of the 24th IEEE International Conference on Advanced Information Networking and Applications (AINA)*, pages 729–736, April 2010a.

Igor Nai Fovino, Marcelo Masera, Luca Guidi, and Giorgio Carpi. An experimental platform for assessing SCADA vulnerabilities and countermeasures in power plants. In *Proceedings of the IEEE International Conference on Human System Interactions (HSI)*, pages 679–686, 2010b.

Igor Nai Fovino, Alessio Coletta, Andrea Carcano, and Marcelo Masera. Critical state-based filtering system for securing SCADA network protocols. *IEEE Transactions on Industrial Electronics*, 59 (10): 3943–3950, October 2012.

A. Frank and A. Asuncion. UCI machine learning repository, 2013a. URL: http://archive.ics .uci.edu/ml.

A. Frank and A. Asuncion. UCI machine learning repository, 2013b. URL: http://archive.ics .uci.edu/ml.

Eibe Frank and Ian Witten. Generating accurate rule sets without global optimization, 1998.

Jerome H. Friedman, Forest Baskett, and Leonard J. Shustek. An algorithm for finding nearest neighbors. *IEEE Transactions on Computers*, 100 (10): 1000–1006, 1975.

Jerome H Friedman, Jon Louis Bentley, and Raphael Ari Finkel. An algorithm for finding best matches in logarithmic expected time. *ACM Transactions on Mathematical Software (TOMS)*, 3 (3): 209–226, 1977.

Keinosuke Fukunaga and Patrenahalli M Narendra. A branch and bound algorithm for computing k-nearest neighbors. *IEEE Transactions on Computers*, 100 (7): 750–753, 1975.

Wei Gao, Thomas Morris, Bradley Reaves, and Drew Richey. On SCADA control system command and response injection and intrusion detection. In *IEEE eCrime Researchers Summit (eCrime)*, pages 1–9, 2010.

Amol Ghoting, Srinivasan Parthasarathy, and Matthew Eric Otey. Fast mining of distance-based outliers in high-dimensional datasets. Springer Journal on *Data Mining and Knowledge Discovery*, 16 (3): 349–364, 2008.

Annarita Giani, Gabor Karsai, Tanya Roosta, Aakash Shah, Bruno Sinopoli, and Jon Wiley. A testbed for secure and robust SCADA systems. *SIGBED Rev*, 5 (2): 1–4, 2008.

Annarita Giani, Eilyan Bitar, Manuel Garcia, Miles McQueen, Pramod Khargonekar, and Kameshwar Poolla. Smart grid data integrity attacks. *IEEE Transactions on Smart Grid*, 4 (3): 1244–1253, Sept 2013.

M. Govindarajan and RM. Chandrasekaran. Evaluation of k-nearest neighbor classifier performance for direct marketing. *Elsevier Journal on Expert Systems with Applications*, 37 (1): 253–258, 2010.

Philip Gross, Janak Parekh, and Gail Kaiser. Secure selecticast for collaborative intrusion detection systems. In *Proceedings of the 3rd International Workshop on Distributed Event-Based Systems (DEBS)*, page 50, 2004.

Siguraur E. Guttormsson, M. A. El-Sharkawi, and I. Kerszenbaum. Elliptical novelty grouping for on-line short-turn detection of excited running rotors. *IEEE Transactions on Energy Conversion (TEC)*, 14: 16–22, 1999.

Mark Hall, Eibe Frank, Geoffrey Holmes, Bernhard Pfahringer, Peter Reutemann, and Ian H. Witten. The weka data mining software J: an update. *ACM SIGKDD Explorations Newsletter*, pages 10–18, 2009.

Trevor Hastie, Robert Tibshirani, Jerome Friedman, T Hastie, J Friedman, and R Tibshirani. *The Elements of Statistical Learning*, volume 2, Springer, 2009.

Mike Hibler, Robert Ricci, Leigh Stoller, Jonathon Duerig, Shashi Guruprasad, Tim Stack, Kirk Webb, and Jay Lepreau. Large-scale virtualization in the emulab network testbed. In *USENIX Annual Technical Conference*, pages 113–128, 2008.

Kevin J. Houle, George M. Weaver, Neil Long, and Rob Thomas. Trends in denial of service attack technology. *CERT Coordination Center*, page 839, 2001.

Allen Householder, Art Manion, Linda Pesante, G Weaver, and Rob Thomas. Managing the threat of denial-of-service attacks. Technical report, CMU Software Engineering Institute CERT Coordination Center, 2001.

Modbus IDA. Modbus messaging on TCP/IP implementation guide v1.0a. June 2004.

Vinay M. Igure, Sean A. Laughter, and Ronald D. Williams. Security issues in SCADA networks. *Elsevier Journal on Computers and Security*, 25 (7): 498–506, 2006.

D. Jablon. Password authentication using multiple servers. In *The Cryptographer's Track at the RSA Conference*, pages 344–360, 2001.

Michael L. Milvich. Idaho national laboratory supervisory control and data acquisition intrusion detection system (SCADA IDS). *In 2008 IEEE Conference on Technologies for Homeland Security*, pages 469–473, 2008.

Meng Jianliang, Shang Haikun, and Bian Ling. The application on intrusion detection based on k-means cluster algorithm. In *Proceedings of International Forum on Information Technology and Applications (IFITA)*, volume 1, pages 150–152, 2009.

Xuan Jin, John Bigham, Julian Rodaway, David Gamez, and Chris Phillips. Anomaly detection in electricity cyber infrastructures. In *Proceedings of the International Workshop on Complex Networks and Infrastructure Protection (CNIP)*, 2006.

George H. John and Pat Langley. Estimating continuous distributions in bayesian classifiers. In *Proceedings of the 11th Conference on Uncertainty in Artificial Intelligence*, pages 338–345. Morgan Kaufmann Publishers Inc., 1995.

I. T. Jolliffe. Principal component analysis and factor analysis. *Principal Component Analysis*, pages 150–166, 2002.

J. Katz, R. Ostrovsky, and M. Yung. Efficient password-authenticated key exchange using human-memorable passwords. In *Proceedings of Eurocrypt*, pages 457–494, 2001.

J. Katz, P. MacKenzie, G. Taban, and V. Gligor. Two-server password-only authenticated key exchange. In *Proceedings of Applied Cryptoraphy and Network Security*, pages 1–16, 2005.

Baek S. Kim and Song B. Park. A fast *k* nearest neighbor finding algorithm based on the ordered partition. *IEEE Transactions on Pattern Analysis and Machine Intelligence (TPAMI)*, (6): 761–766, 1986.

Josef Kittler, Mohamad Hatef, Robert P. W. Duin, and Jiri Matas. On combining classifiers. *IEEE Transactions on Pattern Analysis and Machine Intelligence (TPAMI)*, pages 226–239, 1998.

R. Kohavi. A study of cross-validation and bootstrap for accuracy estimation and model selection. In *Proceedings of International joint Conference on artificial intelligence*, volume 14, pages 1137–1145. Lawrence Erlbaum Associates Ltd, 1995.

Sathish A. P. Kumar, Anup Kumar, and S. Srinivasan. Statistical based intrusion detection framework using six sigma technique. *International Journal of Computer Science and Network Security*, 7 (10): 333, 2007.

Nishchal Kush, Ernest Foo, and Ejaz Ahmed. Smart grid test bed design and implementation. 2010.

A. Lewis. The epanet programmer's toolkit for analysis of water distribution systems, 1999. URL: http://www.epa.gov/nrmrl/wswrd/dw/epanet.html.

Yi-Ching Liaw, Maw-Lin Leou, and Chien-Min Wu. Fast exact k-nearest neighbors search using an orthogonal search tree. *Elsevier Journal on Pattern Recognition*, 43 (6): 2351–2358, 2010.

Ondrej Linda, Todd Vollmer, and Milos Manic. Neural network based intrusion detection system for critical infrastructures. In *Proceedings of the International Joint Conference on Neural Networks (IJCNN)*, pages 1827–1834, June 2009.

Ting Liu, Andrew W Moore, and Alexander Gray. New algorithms for efficient high-dimensional nonparametric classification. *The Journal of Machine Learning Research*, 7: 1135–1158, 2006.

Men Long, Chwan-Hwa Wu, and John Y Hung. Denial of service attacks on network-based control systems: impact and mitigation. *IEEE Transactions on Industrial Informatics (TII)*, 1 (2): 85–96, 2005.

P. MacKenzie, T. Shrimpton, and M. Jakobsson. Threshold password-authenticated key exchange. *Journal of Cryptology*, 19 (1): 27–66, 2006.

J. MacQueen et al. Some methods for classification and analysis of multivariate observations. In *Proceedings of the 5th Berkeley Symposium on Mathematical Statistics and Probability*, volume 1, page 14, California, USA, 1967.

Steen Magnussen, Ronald E. McRoberts, and Erkki O. Tomppo. Model-based mean square error estimators for k-nearest neighbour predictions and applications using remotely sensed data for forest inventories. *Remote Sensing of Environment*, 113 (3): 476–488, 2009.

Matthew V. Mahoney and Philip K. Chan. Learning rules for anomaly detection of hostile network traffic. In *Proceedings the 3rd IEEE International Conference on Data Mining (ICDM)*, pages 601–604, 2003a.

Matthew V. Mahoney and Philip K. Chan. An analysis of the 1999 darpa/lincoln laboratory evaluation data for network anomaly detection. In *Recent Advances in Intrusion Detection*, pages 220–237, Springer, 2003b.

Munir Majdalawieh, Francesco Parisi-Presicce, and Duminda Wijesekera. DNPSec: Distributed network protocol version 3 (DNP3) security framework. In *Advances in Computer, Information, and Systems Sciences, and Engineering*, pages 227–234, 2006.

Brent Martin. Instance-based learning: nearest neighbour with generalisation. PhD thesis, University of Waikato, 1995.

Isabel Marton, Ana I. Sánchezb, Sofia Carlosa, and Sebastián Martorella. Application of data driven methods for condition monitoring maintenance. *Chemical Engineering*, 33: 301–306, 2013.

James McNames. A fast nearest-neighbor algorithm based on a principal axis search tree. *IEEE Transactions on Pattern Analysis and Machine Intelligence (TPAMI)*, 23 (9): 964–976, 2001.

Melbourne Water. Daily residential water use for Melbourne, July 2009. URL: http://www.melbournewater.com.au.

Melbourne Water. Daily residential water use for Melbourne, July 2012. URL: http://www.melbournewater.com.au.

Miljenko Mikuc, Denis Salopek, Valter Vasić, and Marko Zec. The FreeBSD network stack virtualization project, March 2014. URL: http://www.imunes.tel.fer.hr/.

Modbus Organization. Modbus specifications, April 2020. URL: http://www.modbus.org/docs/Modbus_Application_Protocol_V1_1b.pdf (accessed on 21 July 2019).

Thomas Morris, Anurag Srivastava, Bradley Reaves, Wei Gao, Kalyan Pavurapu, and Ram Reddi. A control system testbed to validate critical infrastructure protection concepts. *International Journal of Critical Infrastructure Protection (IJCIP)*, 4 (2): 88–103, 2011.

Kate Munro. Deconstructing flame: the limitations of traditional defences. *Elsevier Journal on Computer Fraud and Security*, 2012 (10): 8–11, 2012.

Gerhard Münz, Sa Li, and Georg Carle. Traffic anomaly detection using k-means clustering. *Proceedings of Leistungs-Zuverlässigkeits-und Verlässlichkeitsbewertung von Kommunikationsnetzen und Verteilten Systemen*, 4, 2007.

Sameer A. Nene and Shree K. Nayar. A simple algorithm for nearest neighbor search in high dimensions. *IEEE Transactions on Pattern Analysis and Machine Intelligence (TPAMI)*, 19 (9): 989–1003, 1997.

Peng Ning, Yun Cui, and Douglas S. Reeves. Constructing attack scenarios through correlation of intrusion alerts. In *Proceedings of the 9th ACM Conference on Computer and Communications Security (CCS)*, pages 245–254, November 2002.

NS3 Maintainers. Ns3 simulator, July 2012. URL: http://www.nsnam.org/.

Joshua Oldmeadow, Siddarth Ravinutala, and Christopher Leckie. Adaptive clustering for network intrusion detection. *Springer Journal on Advances in Knowledge Discovery and Data Mining*, pages 255–259, 2004.

Paul Oman, Edmund Schweitzer, and Deborah Frincke. Concerns about intrusions into remotely accessible substation controllers and SCADA systems. In *Proceedings of the Twenty-Seventh Annual Western Protective Relay Conference*, volume 160. Citeseer, 2000.

Stephen M. Omohundro. *Five Balltree Construction Algorithms*. International Computer Science Institute Berkeley, 1989.

OPNET Technologies. Opnet modeler: Scalable network simulation, July 2012. URL: http://www.opnet.com/solutions/network_rd/modeler.html.

Apostolos N. Papadopoulos. *Nearest Neighbor Search: A Database Perspective*. Springer, 2006.

Al-Sakib Khan Pathan. *The State of the Art in Intrusion Prevention and Detection*. CRC Press, 2014.

Roberto Di Pietro and Luigi V. Mancini. *Intrusion Dtection Systems*. Springer, 2008.

PLeonid Portnoy, Eleazar Eskin, and Sal Stolfo. Intrusion detection with unlabeled data using clustering. 2001.

Michael J. Prerau and Eleazar Eskin. Unsupervised anomaly detection using an optimized k-nearest neighbors algorithm. *Undergraduate Thesis*, Columbia University, December 2000.

Matija Pužar and Thomas Plagemann. NEMAN: A network emulator for mobile ad-hoc networks. Technical report. University of Oslo, 2005.

Carlos Queiroz, Abdun Mahmood, and Zahir Tari. SCADASim - a framework for building SCADA simulations. *IEEE Transactions on Smart Grid (TSG)*, 2 (4): 589–597, 2011.

John Ross Quinlan. *C4.5: Programs for Machine Learning*, volume 1. Morgan Kaufmann, 1993.

M. Di Raimondo and R. Gennaro. Provably secure threshold password-authenticated key exchange. *Journal of Computer and System Sciences*, 72 (6): 978–1001, 2006.

Bradley Reaves and Thomas Morris. An open virtual testbed for industrial control system security research. *Springer International Journal of Information Security*, pages 1–15, 2012.

Irene Rodriguez-Lujan, Jordi Fonollosa, Alexander Vergara, Margie Homer, and Ramon Huerta. On the calibration of sensor arrays for pattern recognition using the minimal number of experiments. *Elsevier Journal on Chemometrics and Intelligent Laboratory Systems*, pages 123–134, 2014.

J. L. Rrushi, C. Bellettini, and E. Damiani. *Composite intrusion detection in process control networks*. PhD thesis, University of Milano, April 2009a.

J. L. Rrushi, C. Bellettini, and E. Damiani. *Composite intrusion detection in process control networks*. PhD thesis, University of Milano, April 2009b.

Sandia National Laboratories. National SCADA testbed, July 2012. URL: http://energy.sandia.gov/?page_id=859.

Salvatore Sanfilippo. Hping3-active network security tool, 2012. URL: http://www.hping.org/hping3.html.

Bernhard Schiilköpf, Chris Burgest, and Vladimir Vapnik. Extracting support data for a given task. In *Proceedings of the 1st International Conference on Knowledge Discovery and Data Mining, AAAI Press*, pages 252–257, 1995.

Gregory Shakhnarovich, Trevor Darrell, and Piotr Indyk. Nearest-neighbor methods in learning and vision. *IEEE Transactions on Neural Networks (TNN)*, 19 (2): 377, 2008.

A. Shamir. How to share a secret key. *Communications of the ACM (CACM)*, 22 (11): 612–613, 1979.

Haijian Shi. *Best-first decision tree learning*. PhD thesis, Citeseer, 2007.

Almas Shintemirov, Wenhu Tang, and Q. H. Wu. Power transformer fault classification based on dissolved gas analysis by implementing bootstrap and genetic programming. *IEEE Transactions on Systems, Man, and Cybernetics, Part C: Applications and Reviews*, 39 (1): 69–79, 2009.

PowerWorld Simulator. Power generation simulator, July 2013. URL: http://www.powerworld.com.

Jill Slay and Michael Miller. *Lessons Learned from the Maroochy Water Breach*. Springer, 2007.

Software Development Team. Advanced hmi development software package, November 2014. URL URL: http://http://advancedhmi.com/.

Robert F. Sproull. Refinements to nearest-neighbor searching ink-dimensional trees. *Algorithmica*, 6 (1-6): 579–589, 1991.

Shan Suthaharan, Mohammed Alzahrani, Sutharshan Rajasegarar, Christopher Leckie, and Marimuthu Palaniswami. Labelled data collection for anomaly detection in wireless sensor networks. In *Proceedings of the 6th International Conference on Intelligent Sensors, Sensor Networks and Information Processing (ISSNIP)*, pages 269 –274, Dec 2010.

Modbus Team. Modbus implementation and tools, July 2014. URL: https://github.com/bashwork/pymodbus.

Patrick P. Tsang and Sean W. Smith. Yasir: A low-latency, high-integrity security retrofit for legacy SCADA systems. In *Proceedings of The Ifip Tc 11 23rd International Information Security Conference*, pages 445–459. Springer, 2008.

TU Berlin. Simulation of wind turbine blades, October 2013. URL: http://q-blade.org/.

Alfonso Valdes and Steven Cheung. Communication pattern anomaly detection in process control systems. In *Proceedings of IEEE International Conference on Technologies for Homeland Securitym (HST)*, pages 22–29, 2009.

Andras Varga and Rudolf Hornig. An overview of the OMNeT++ simulation environment. In *Proceedings of the 1st International ConfeNetworks and Systems and Workshops*, page 60. ICST (Institute for Computer Sciences, Social-Informatics and Telecommunications Engineering), 2008.

Alexander Vergara, Shankar Vembu, Tuba Ayhan, Margaret A. Ryan, Margie L. Homer, and Ramon Huerta. Chemical gas sensor drift compensation using classifier ensembles. *Elsevier Journal on Sensors and Actuators B: Chemical*, 166: 320–329, 2012.

Šaltenis Vydunas. Outlier detection based on the distribution of distances between data points. *Informatica*, 15 (3): 399–410, August 2004.

Dong Wei, Yan Lu, Mohsen Jafari, Paul M. Skare, and Kenneth Rohde. Protecting smart grid automation systems against cyberattacks. *IEEE Transactions on Smart Grid (TSG)*, (99): 1–1, 2011.

Fangfei Weng, Qingshan Jiang, Liang Shi, and Nannan Wu. An intrusion detection system based on the clustering ensemble. In *Proceedings of IEEE International Workshop on Anti-counterfeiting, Security, Identification*, pages 121–124, 2007.

Yang Wenxian and Jiang Jiesheng. Wind turbine condition monitoring and reliability analysis by SCADA information. In *Proceedings of the 2nd International Conference on Mechanic Automation and Control Engineering (MACE)*, pages 1872–1875, July 2011.

Xindong Wu, Vipin Kumar, J. Ross Quinlan, Joydeep Ghosh, Qiang Yang, Hiroshi Motoda, Geoffrey J. McLachlan, Angus Ng, Bing Liu, Philip S. Yu, Zhi-Hua Zhou, Michael Steinbach, David J. Hand, and Dan Steinberg. Top 10 algorithms in data mining. *Elsevier Journal on Knowledge and Information Systems*, 14 (1): 1–37, 2008.

Wang Xueyi. A fast exact k-nearest neighbors algorithm for high dimensional search using k-means clustering and triangle inequality. In *IJCNN Proceedings*, pages 1293–1299, 2011.

Kenji Yamanishi and Jun ichi Takeuchi. Discovering outlier filtering rules from unlabeled data. In *Proceedings of inte Knowledge Discovery and Data Mining*, pages 389–394. Citeseer, 2001.

Dayu Yang, Alexander Usynin, and J. Wesley Hines. Anomaly-based intrusion detection for SCADA systems. In *Proceedings of the 5th Intl. Topical Meeting on Nuclear Plant Instrumentation, Control and Human Machine Interface Technologies (NPIC&HMIT)*, pages 12–16, 2006.

X. Yi, S. Ling, and H. Wang. Efficient two-server password-only authenticated key exchange. *IEEE Transactions on Parallel Distributed Systems (TPDS)*, 24 (9): 1773–1782, 2013.

X. Yi, F. Hao, and E. Bertino. Id-based two-server password-authenticated key exchange. In *Proceedings of European Symposium on Research in Computer Security (ESORICS)*, pages 257–276, 2014.

X. Yi, F. Hao, L. Chen, and J. K. Liu. Practical threshold password-authenticated secret sharing protocol. In *Proceedings of European Symposium on Research in Computer Security (ESORICS)*, pages 347–365, 2015.

A. S. A. E. Zaher, S. D. J. McArthur, D. G. Infield, and Y. Patel. Online wind turbine fault detection through automated SCADA data analysis. *Wind Energy*, 12 (6): 574–593, 2009.

Ji Zhang and Hai Wang. Detecting outlying subspaces for high-dimensional data: The new task, algorithms, and performance. *Springer Journal on Knowledge and Information Systems*, 10 (3): 333–355, 2006.

Jun Zhang, Chao Chen, Yang Xiang, Wanlei Zhou, and Yong Xiang. Internet traffic classification by aggregating correlated naive bayes predictions. *IEEE Transactions on Information Forensics and Security (TIFS)*, 8 (1): 5–15, 2013a.

Jun Zhang, Yang Xiang, Yu Wang, Wanlei Zhou, Yong Xiang, and Yong Guan. Network traffic classification using correlation information. *IEEE Transactions on Parallel and Distributed Systems (TPDS)*, 8 (1): 104–117, 2013b.

INDEX

SCADA Security: Machine Learning Concepts for Intrusion Detection and Prevention,
First Edition. Abdulmohsen Almalawi, Zahir Tari, Adil Fahad and Xun Yi.
© 2021 John Wiley & Sons, Inc. Published 2021 by John Wiley & Sons, Inc.

Wiley Series on Parallel and Distributed Computing
Series Editor: Albert Y. Zomaya

Parallel and Distributed Simulation Systems
Richard Fujimoto

Mobile Processing in Distributed and Open Environments
Peter Sapaty

Introduction to Parallel Algorithms
C. Xavier and S. S. Iyengar

Solutions to Parallel and Distributed Computing Problems: Lessons from Biological Sciences
Albert Y. Zomaya, Fikret Ercal, and Stephan Olariu (Editors)

Parallel and Distributed Computing: A Survey of Models, Paradigms, and Approaches
Claudia Leopold

Fundamentals of Distributed Object Systems: A CORBA Perspective
Zahir Tari and Omran Bukhres

Pipelined Processor Farms: Structured Design for Embedded Parallel Systems
Martin Fleury and Andrew Downton

Handbook of Wireless Networks and Mobile Computing
Ivan Stojmenović (Editor)

Internet-Based Workflow Management: Toward a SemanticWeb
Dan C. Marinescu

Parallel Computing on Heterogeneous Networks
Alexey L. Lastovetsky

Performance Evaluation and Characterization of Parallel and Distributed Computing Tools
Salim Hariri and Manish Parashar

Distributed Computing: Fundamentals, Simulations, and Advanced Topics, Second Edition
Hagit Attiya and JenniferWelch

Smart Environments: Technology, Protocols, and Applications
Diane Cook and Sajal Das

Fundamentals of Computer Organization and Architecture
Mostafa Abd-El-Barr and Hesham El-Rewini

Advanced Computer Architecture and Parallel Processing
Hesham El-Rewini and Mostafa Abd-El-Barr

UPC: Distributed Shared Memory Programming
Tarek El-Ghazawi,William Carlson, Thomas Sterling, and Katherine Yelick

Handbook of Sensor Networks: Algorithms and Architectures
Ivan Stojmenović (Editor)

Parallel Metaheuristics: A New Class of Algorithms
Enrique Alba (Editor)

Design and Analysis of Distributed Algorithms
Nicola Santoro

Task Scheduling for Parallel Systems
Oliver Sinnen

Computing for Numerical Methods Using Visual C++
Shaharuddin Salleh, Albert Y. Zomaya, and Sakhinah A. Bakar

Architecture-Independent Programming for Wireless Sensor Networks
Amol B. Bakshi and Viktor K. Prasanna

High-Performance Parallel Database Processing and Grid Databases
David Taniar, Clement Leung,Wenny Rahayu, and Sushant Goel

Algorithms and Protocols for Wireless and Mobile Ad Hoc Networks
Azzedine Boukerche (Editor)

Algorithms and Protocols for Wireless Sensor Networks
Azzedine Boukerche (Editor)

Optimization Techniques for Solving Complex Problems
Enrique Alba, Christian Blum, Pedro Isasi, Coromoto León, and Juan Antonio Gómez (Editors)

Emerging Wireless LANs, Wireless PANs, and Wireless MANs: IEEE 802.11, IEEE 802.15, IEEE 802.16 Wireless Standard Family
Yang Xiao and Yi Pan (Editors)

High-Performance Heterogeneous Computing
Alexey L. Lastovetsky and Jack Dongarra

Mobile Intelligence
Laurence T. Yang, Augustinus Borgy Waluyo, Jianhua Ma, Ling Tan, and Bala Srinivasan (Editors)

Research in Mobile Intelligence
Laurence T. Yang (Editor)

Advanced Computational Infrastructures for Parallel and Distributed Adaptive Applications
Manish Parashar and Xiaolin Li (Editors)

Market-Oriented Grid and Utility Computing
Rajkumar Buyya and Kris Bubendorfer (Editors)

Cloud Computing Principles and Paradigms
Rajkumar Buyya, James Broberg, and Andrzej Goscinski (Editors)

Algorithms and Parallel Computing
Fayez Gebali

Energy-Efficient Distributed Computing Systems
Albert Y. Zomaya and Young Choon Lee (Editors)

Scalable Computing and Communications: Theory and Practice
Samee U. Khan, LizheWang, and Albert Y. Zomaya (Editors)

The DATA Bonanza: Improving Knowledge Discovery in Science, Engineering, and Business
Malcolm Atkinson, Rob Baxter, Michelle Galea, Mark Parsons, Peter Brezany, Oscar Corcho, Jano van Hemert, and David Snelling (Editors)

Large Scale Network-Centric Distributed Systems
Hamid Sarbazi-Azad and Albert Y. Zomaya (Editors)